FINISH CARPENTRY

FROM THE EDITORS OF **Fine Homebuilding**®

The Taunton Press

The Taunton Press, Inc., 63 South Main Street, PO Box 5506, Newtown, CT 06470-5506
e-mail: tp@taunton.com

Distributed by Publishers Group West

DESIGN AND LAYOUT: Cathy Cassidy

COVER PHOTOGRAPHERS: Front Cover: Charles Miller, courtesy *Fine Homebuilding,* © The Taunton Press, Inc.; Back cover: © Patrick Cudahy (top left); Rich Ziegner, courtesy *Fine Homebuilding,* © The Taunton Press, Inc. (top right); Charles Miller, courtesy *Fine Homebuilding,* © The Taunton Press, Inc. (bottom left, bottom right)

Fine Homebuilding® is a trademark of The Taunton Press, Inc., registered in the U.S. Patent and Trademark Office

The following manufacturers/names appearing in *Finish Carpentry* are trademarks: Airy®; Andersen℠; Black & Decker®, Inc.; Bosch®; Cross®; Delta®; Duo-Fast®; Elu®; Fasco™; Freud®; Hitachi®; Hilti®; Kaiser®; Lamello®; Lexan®; Lubriplate®; Paslode®; Porter-Cable®; Ryobi®; Senco®; Skil®; Stanley®-Bostitch®; Stevens Institute of Technology®; 3M®; Virutex™; WD-40®; Wetordry® TRI-M-ITE®.

Finish carpentry / editors of Fine homebuilding
        p. cm. -- (Taunton's for pros/by pros)
Includes index.
  ISBN 1-56158-536-X
  1. Finish carpentry. I. Fine homebuilding. II. For pros, by pros.
TH5640 .F56 2002
694' .6--dc21   2002010193

Printed in United States of America
10 9 8 7 6 5 4 3 2 1

Special thanks to the authors, editors, art directors,

copy editors, and other staff members of *Fine Homebuilding*

who contributed to the development of the articles in this book.

CONTENTS

PART 4: CROWN MOLDING

PART 5: PANELING

INTRODUCTION

"That gap's so big you could throw a cat through it." I had just finished trimming a window, when one of my fellow carpenters assessed the quality of a particular miter joint with that comment. It was job-site hyperbole, of course. You could barely have slipped a matchbook cover into the gap, but his point was clear. The gap was too big.

Finish carpentry is not a game of inches. It's a game of skoshes, hairs, tads, and other increments smaller than any carpenter's tape will measure. The overriding goal of frame carpentry is strength, but with finish carpentry, it's all about looks. And in order to look good, finish carpentry must be executed to very high tolerances.

In the real world, working to high tolerances is hard. Floors are never level, walls are never plumb, and the build-up of joint compound on drywall means that corners are never square. There's never a stud or joist to nail into when you need one. And then there's human nature to contend with. You know, the voice whispering in your ear that you don't really need to drill a pilot hole for that nail, which of course splits the wood as soon as you drive home the nail.

To do good finish work you need four things. First you need to care about doing good work. Then you need patience and good tools. And finally you need every trick in the book . . . which makes this a good place to start. This book is actually a collection of articles orginally published in *Fine Homebuilding* magazine. Written by builders from all over the country, these articles contain the hard-won lessons from their real-world experience.

—Kevin Ireton,
editor-in-chief, *Fine Homebuilding*

Basic Scribing Techniques

■ BY JIM TOLPIN

The first thing I learned as a finish carpenter was that square corners, plumb walls, and level floors and ceilings don't exist on this planet. And because that's just the way it is, it was up to me to learn how to work with these unfortunate divergencies from the way it ought to be. As the finish man, my job was to fit the pretty stuff to the structures that framers and rockers left behind, no matter how crooked they were.

In my quest for perfect fits, I learned how to use bevel squares and base hooks, among other tools, and became proficient in the use of a slightly customized pencil compass. I learned from legendary boat builder Bud MacIntosh how to use something called a spiling batten to solve certain awkward scribing problems, such as fitting the last ceiling board. I even paid homage to the linoleum trade and learned the ingeniously simple "Joe Frogger" method of creating a template that can produce dead-accurate fits every time.

Using the Bevel Square

A bevel square is a layout tool with a wood, metal, or plastic body having an adjustable metal blade attached to one end. The square is used mostly for determining the angle at which a piece of trim needs to be cut to fit tightly against a surface.

My first bevel square came from my grandfather. It's a nice rosewood-bodied job with a 6-in.-long blade. It's pretty and has sentimental value, but like many contemporary bevel squares, it's not the best tool for taking angles. This is because its locking lever, which is located at the pivot point of the tool, often sticks beyond the edge of the body and gets in the way. Also, the body is quite thick, which holds the blade away from the stock. This can throw off the angle measurement. What's more, the body is relatively short, which can also produce inaccurate readings.

I like my all-metal Japanese bevel square better. It's much thinner than a conventional bevel square; the lock is a knurled knob that's out of the way; and it can be held and locked with one hand.

Although the use of a bevel square may seem straightforward, it's not. Always extend the blade fully before pressing the outside edge of the body against a surface to measure an angle (such as when measuring an inside corner where two walls meet). Any protrusion of the blade beyond the outside

Fitting Baseboard

Author Jim Tolpin uses either a bevel square (top photo) or a homemade base hook (bottom photo) to lay out baseboards for a tight fit against door casings. Bumpy floors can fool a bevel square, so always measure the bevel off of a straight-edge or a piece of the baseboard itself. The base hook is used by holding it hard against the casing (or plinth block in this case) while scribing a cutline directly on the baseboard.

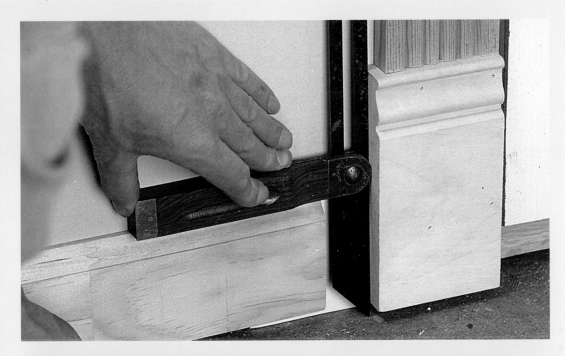

edge of the body will hold the body away from the surface it's resting against, throwing off the angle reading.

Also, don't assume that you can simply press the square against converging surfaces to get an accurate reading. Say, for instance, that you want to fit a baseboard to a door casing (top photo, p. 5). To measure the angle of the end cut, set the baseboard where you want it on the floor, then place the body of your bevel square on top of the baseboard to measure the angle of the casing. If you simply lay the body of the square on the floor, any bumps or dips in the floor next to the joint will fool the square into measuring a false angle. An alternative is to set a level or a straightedge on the floor and to measure the angle off of that.

Once you measure an angle, be careful not to jar the bevel before you scribe the workpiece. Fortunately, there's an easy way to ensure against the loss of an angle setting on a bevel: Record it with the help of a boat builder's bevel board.

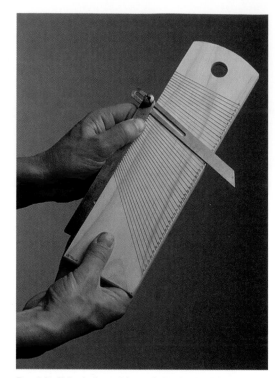

Boat builder's bevel board. Etched with 46 labeled lines spaced 1° apart, the homemade bevel board makes it easy to read an angle off a bevel square, then adjust a saw to that angle.

TIP

If you don't have a bevel board, scribe and label each angle on a wood block right after you measure it, then reset your bevel square from the block to mark your trim.

Saving Angles

Boat builders, who confront compound angles on nearly every piece they fit, have developed a simple, shopmade accessory that makes it easy to measure and record a series of angles for future reference at the saw table. Called a bevel board, it's a board with a bunch of lines drawn across it at angles ranging from 0° to 45° (photo, above). The bevel board allows you to measure an angle with your bevel square and then read the degrees of the angle directly from the board. If the angle scale on your bandsaw, table saw, or chopsaw is calibrated to the bevel board, you need only to set the saw to the appropriate degree mark and cut away. If more than one angle is being taken at once, the angles are simply recorded on a scrap of wood or paper that represents a story board of the piece or pieces to be cut.

The bevel board should be made of a stable wood, such as mahogany or teak, that has an interlocking, split-resistant grain. You can also use a scrap of ⅜-in. plywood. Using a protractor, scribe the lines to the board with an awl and then fill them in with an indelible ink. Keep the board thin so that it will be lightweight (⅜ in. thick is sufficient); leave room at both ends of the board for indexing the body of the bevel against it; and radius or chamfer the top edge of the board so that you can orient it at a glance. An alternative board, sans the romance (and not quite as easy to read), is made by scratching the lines deeply into a piece of Lexan® plastic.

In lieu of a bevel board, you can scribe and label each angle on a wood block right after you measure it, then reset your bevel square from the block to mark your trim. If you need to quantify an angle in degrees, measure it on the block with a Speed Square.

The Amazing Pencil Compass

It's amazing what the pencil compass allows you to do. For instance, it really comes into its own for fitting a wall panel or a vertical siding board to a bumpy surface, such as a fireplace. The procedure is straightforward. First, plumb the panel or board and tack it to the wall about ½ in. away from the closest spot on the meeting surface. Then set the compass to distance "X" between the edge of the panel and the bottom of the deepest valley on the meeting surface, plus ¼ in. so that the scribed line won't fall off the edge of the workpiece. Hold the compass level along the entire vertical run and trace the meeting surface with the feeling point so that the pencil transfers the profile to the workpiece. (If the workpiece is dark, a strip of wide painter's masking tape applied to the panel will make the line more legible.) Finally, re-move the workpiece from the wall, back-cut it (bevel it back) a few degrees along the scribed line, then test fit it against the meeting surface. If the fit is good in some areas and way off in others, you probably let the compass wander from level during scribing. If this happens, try again. If neces-sary, final fitting is achieved through hand-planing, sanding, rasping, and filing (more on that later).

Scribing a Panel

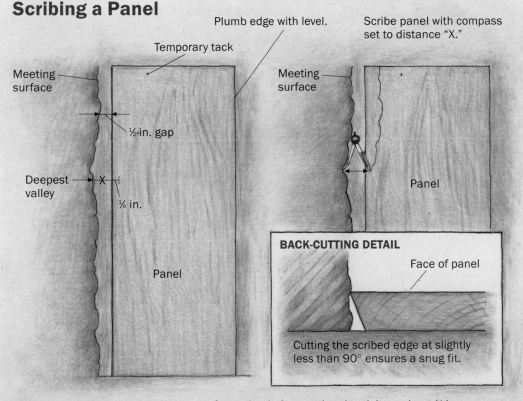

To scribe a wall panel to an uneven surface, plumb the panel and tack it up about ½ in. away from the closest spot on the meeting surface. Then scribe the panel with the compass points set to distance "X" (the distance between the edge of the panel and the bottom of the deepest valley on the meeting surface plus ¼ in.). Back-cutting the panel will ensure a snug fit.

Scribing the Closing Board

The installation of wainscoting and the installation of vertical siding both have the same problem—fitting the closing (or last) board. I approach this problem by nailing up all but the last few boards. Then I tack up the rest of the boards except for the last one. I mark and cut this closing board, remove the tacked ones, then spring the whole group into place at once and nail them to the wall.

How I fit the closing board depends on the nature of the surface it meets. If the meeting surface is straight and plumb, I simply measure the gap and rip the board to width. If the meeting surface is straight but not plumb, I measure across the top and bottom of the gap, transfer the measurements to the board, connect the marks with a straight line, and rip the board with a circular saw.

If the meeting surface is irregular, the board needs to be scribed as shown in the drawings at right. In this case, after I've tacked up the next-to-last board, I mark the top and bottom of its leading edge on the underlying wall (points A and B). Then I remove all of the tacked-up boards, hold the closing board hard against the meeting surface, and mark the top and bottom of its trailing edge on the wall (points C and D).

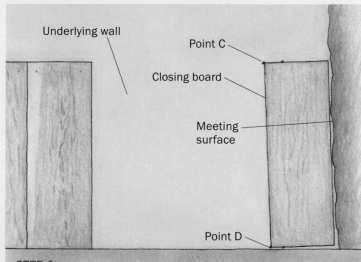

Here's a four-step method for fitting the last vertical board on a wall to a bumpy surface.

Last few boards tacked temporarily to wall.

Point A

Gap to be filled by last board.

Point B

Underlying wall

STEP 1:
Install all but the last board on the wall, tacking up the last few boards for easy removal. Mark the leading edge of the second-to-last board on the wall (points A and B).

Underlying wall

Point C

Closing board

Meeting surface

Point D

STEP 2:
Remove the tacked-up boards, hold the closing board hard against the meeting surface, and mark the top and the bottom of the board along its trailing edge (points C and D).

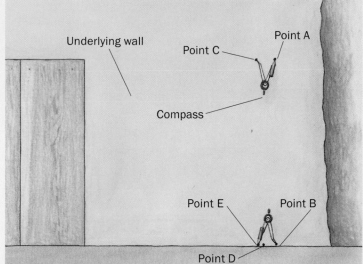

Underlying wall

Point A

Point C

Compass

Point E

Point B

Point D

STEP 3:
Adjust a pencil compass to span either the top or the bottom two marks, whichever are the farthest apart (points A and C this time). Use the compass at this setting to mark point E on the wall.

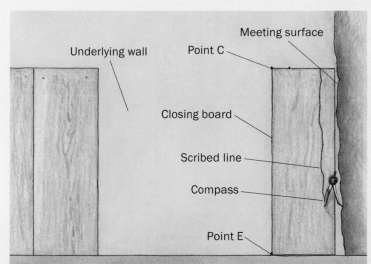

Meeting surface

Underlying wall

Point C

Closing board

Scribed line

Compass

Point E

STEP 4:
Align the trailing edge of the closing board with points C and E, then scribe the board off of the meeting surface with the compass setting unchanged. Once the board is cut to fit, spring it and the remaining boards in place and nail them to the wall.

Setting the closing board aside, I now adjust my compass to span either the top or the bottom two marks, whichever are the farthest apart (points A and C in the example). I then use the compass to make a new mark (point E) at the opposite end so that the top and bottom pairs of marks are the same distance apart (if they weren't to begin with). Finally, I replace the closing board so that its trailing edge falls on the appropriate marks (C and E), then use the compass, with its setting unchanged, to scribe the board to the meeting surface.

I almost always make the cut with a handsaw, cleaning up to the line with a block plane if necessary. A handsaw cuts on the downstroke, ensuring that any tearout will occur on the backside of the board where it won't show. I undercut the board slightly so that when it's sprung into place it makes a nice, neat joint.

When installing wall paneling instead of boards, only the second-to-last panel is tacked up and removed to fit the last piece. Otherwise, the scribing and cutting procedure is the same.

Tom Law is a former consulting editor for Fine Homebuilding. He *lives in Smithsburg, Maryland.*

The Base Hook

Another homemade tool, called a base hook, eliminates the need for a bevel square in some applications. Similar in concept to a siding gauge, it's simply an L-shaped piece of a stable, split-resistant wood used primarily for laying out the end cut of baseboard where it butts against standing moldings such as door casings (bottom photo, p. 5). To use the hook, lap it over the baseboard and hold it hard against the standing molding while scribing a cutline across the baseboard. Be sure the faces of your base hook are perfectly square to the edges, or you'll introduce a margin of error.

Scribing to Irregular Surfaces

Shortly after I became a finish carpenter, I bought a $5 pencil compass like the kind my kids tote in their school bags. It has two adjustable arms, with a metal feeler point at the end of one arm and a pencil at the end of the other. For improved accuracy, I heated and bent out the feeler point of my compass slightly so that the point, rather than a portion of its side, contacts the meeting surface. (The meeting surface is whatever is being scribed to; I'll call the piece to be cut the workpiece.)

Scribing with a Compass

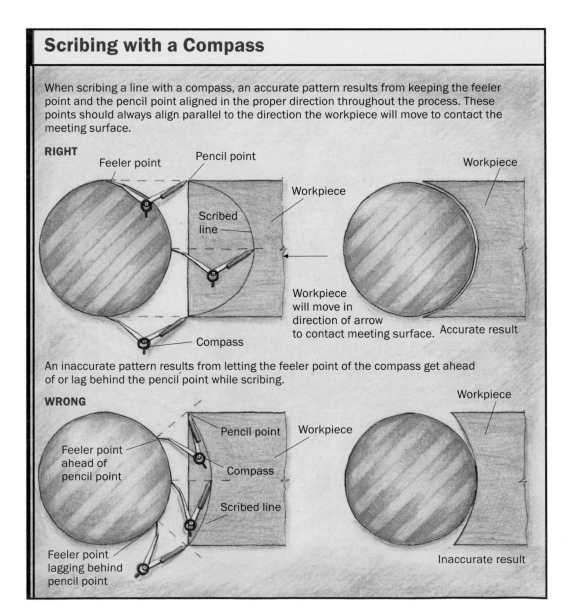

When scribing a line with a compass, an accurate pattern results from keeping the feeler point and the pencil point aligned in the proper direction throughout the process. These points should always align parallel to the direction the workpiece will move to contact the meeting surface.

RIGHT

Feeler point

Pencil point

Workpiece

Scribed line

Workpiece

Compass

Workpiece will move in direction of arrow to contact meeting surface.

Workpiece

Accurate result

An inaccurate pattern results from letting the feeler point of the compass get ahead of or lag behind the pencil point while scribing.

WRONG

Pencil point

Workpiece

Feeler point ahead of pencil point

Compass

Scribed line

Feeler point lagging behind pencil point

Workpiece

Workpiece

Inaccurate result

Although I haven't tried it yet, I recently learned a tip from Gary Katz, a contractor in California. To ensure that he can always scribe a fine line, Katz fits his compass with a Cross® #3503 mechanical pencil. This pencil is expensive ($15.50), but it scribes a very fine line, is well made, and has a wonderful warranty. No matter how you damage it and regardless of its age, you can return it to Cross, and they'll send you a new one.

When scribing a line with a compass, you are actually transferring the pattern of the meeting surface onto the workpiece. It is very important, as you scribe the line, that the feeler point on one side of the compass not get ahead of or lag behind the pencil point on the other side. Throughout the scribing process, these two points must align parallel to the direction the workpiece will move to contact the meeting surface. If they don't, the result will be an inaccurate pattern and, ultimately, a sloppy fit (drawing, facing page). Chinkless-log-home builders, who routinely scribe logs to fit together tightly, have developed a homemade compass with an adjust-able bubble level on it that makes it easy to keep the compass oriented properly while scribing.

Sometimes the closing (last) board or panel on a wall must be scribed. This is tougher to do because the board has to fit into an existing gap. For one solution to this problem, see Tom Law's method on pp. 8–9.

Another common scribing problem is fitting stair treads between a pair of skirtboards (or similarly, closet shelves between two walls). This is accomplished by cutting the tread ¾ in. longer than its final length; dropping it into place with one end riding high on a skirtboard; scribing and cutting the low end; marking the final length of the tread by measuring off the scribed end; dropping the tread back into place with its scribed end riding high; and then scribing and cutting the low end to the measurement mark.

Fitting a Window Stool

The author uses a combination square to mark the outside corner of the window opening on the stool (top photo). Then he uses a bevel square to measure the angle between the window jamb and the front edge of the stool. This angle is scribed on the stool through the corner point (middle photo). Finally, he sets his pencil compass to span the distance between the corner of the opening and the corner mark on the stool, then scribes the stool's horn to the wall (bottom photo).

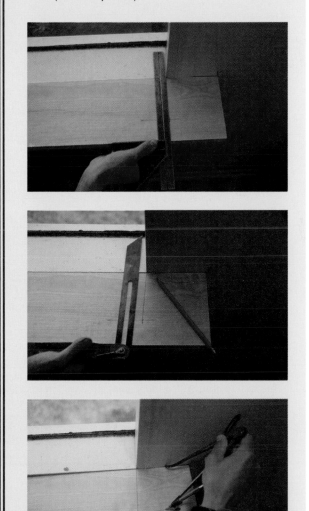

Fitting the Last Ceiling Board
(Looking up at the Ceiling)

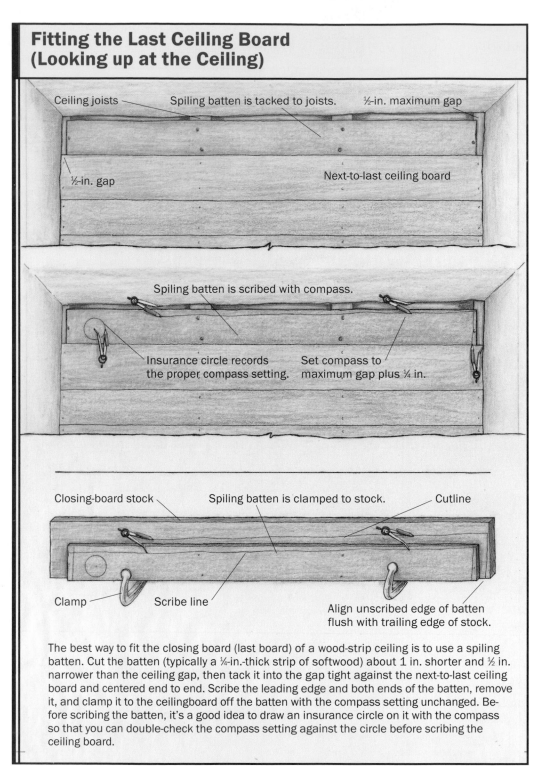

Ceiling joists   Spiling batten is tacked to joists.   ½-in. maximum gap

½-in. gap

Next-to-last ceiling board

Spiling batten is scribed with compass.

Insurance circle records the proper compass setting.

Set compass to maximum gap plus ¼ in.

Closing-board stock   Spiling batten is clamped to stock.   Cutline

Clamp   Scribe line

Align unscribed edge of batten flush with trailing edge of stock.

The best way to fit the closing board (last board) of a wood-strip ceiling is to use a spiling batten. Cut the batten (typically a ¼-in.-thick strip of softwood) about 1 in. shorter and ½ in. narrower than the ceiling gap, then tack it into the gap tight against the next-to-last ceiling board and centered end to end. Scribe the leading edge and both ends of the batten, remove it, and clamp it to the ceilingboard off the batten with the compass setting unchanged. Before scribing the batten, it's a good idea to draw an insurance circle on it with the compass so that you can double-check the compass setting against the circle before scribing the ceiling board.

window opening (where the wall meets the side jambs); a bevel square to lay out the angles of the side jambs relative to the front edge of the stool; and a compass to scribe the stool horns to the wall. For added convenience, a couple of sticks tacked to the sill (perpendicular to the window) will support the stool while you lay it out.

Scribing with a Spiling Batten

Sometimes it's awkward to hold a workpiece in position for scribing. A perfect example is scribing the closing board of a wood-strip ceiling. Not only do you have to hold up the board while scribing it, but the oversize board tilts into the opening. This tilt can throw off the scribed line.

Boat builders confront this exact situation when planking a wooden hull, and they've come up with a nifty device to cope with it: the spiling batten. The spiling batten is simply a thin strip of wood (¼-in.-thick softwood is standard) that's tacked into the opening that the last plank (or shutter) will have to fill. The batten is scribed (or spiled, as boat builders would say) to the meeting surfaces along its leading edges and ends, then removed and clamped to the workpiece. The scribe is then reproduced in reverse, from the batten back to the work.

Some fitting jobs are accomplished using a pencil compass in concert with a bevel square and a combination square. Laying out a window stool is a good example (sidebar, p. 11). In this case, use a combination square to locate the outside corners of the

For ceilings, I cut the batten about 1 in. shorter and ½-in. narrower than the ceiling gap, then tack it up snug against the next-to-last ceiling board, spaced ½ in. shy of the wall at either end (drawings, facing page). I then set my compass to the maximum gap between the batten and the wall, plus ¼ in. to make sure that the scribed line doesn't veer off the edge of the batten. Before scribing the batten, I draw a circle on the batten with the compass to serve as a reference if the compass is bumped inadvertently. Once the batten is scribed, I remove it and clamp it to the closing board, positioned with its trailing edge (the edge that meets the second-to-last ceiling board) flush with the trailing edge of the board. The board is then scribed off the batten with the compass setting unchanged (double-checked against the insurance circle on the batten).

I back-cut the ceiling board about 5° to allow the board to swing into place. And this cut makes it easy to plane the board to fit if necessary.

Fitting Floors to Posts

The bottom photo on p. 14 shows a wide floorboard that fits tightly around a post. If the post had been square and its faces flat, I would have laid out the floorboard using a combination square. But, of course, the post isn't perfect, and the combination square stayed in the toolbox.

Instead, I called on the Joe Frogger method, as it's known in the linoleum trade, to make a template that works like a spiling batten. You'll need a pencil, a utility knife, a piece of heavy felt paper or noncorrugated cardboard for the template, and a small block of wood measuring about 1½ in. square by about ½ in. thick (the block is the frog).

The procedure is simple. First, use the utility knife to cut an opening in the template that matches the profile of the post, adding about ¾ in. clearance all around. Slip the template around the post, tight against the last installed floorboard, and attach it to

A trick of the linoleum trade called "Joe Frogger" makes it easy to fit a floorboard to a post. First, cut a cardboard template to fit around the post and tape it to the subfloor, tight against the previously installed floorboard. Next, hold a small block of wood (the frog) against the post in several spots while marking the frog's outside edge on the template. (Continued on p. 14.)

After removing the template, tape it over the next floorboard and use the frog to transfer each mark from the template to the board.

The marks are joined using a straightedge and a pencil. The reward is a perfect fit.

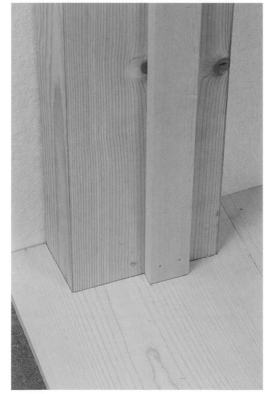

the subfloor with double-stick tape. Then hold the frog against the post at stations spaced a couple of inches apart while you mark along the outside edge of the frog on the template with a sharp pencil. Rabbets at opposite ends of the frog make it easy to orient the frog in the same way at each station (scribing is always done off a rabbeted edge).

Next, remove the marked template from the subfloor and tape it to the floorboard to be fit, flush with the board's end and trailing edge (photo, left). Then index a rabbeted end of the frog against each mark on the template while you mark the opposite end on the floorboard. Finally, join the marks using a pencil and a straightedge, then back-cut slightly along the cutlines. If you're careful, you'll be rewarded with a perfect fit.

Cutting It

Once a workpiece is laid out, there are a number of ways to cut it. Unless the cutlines are relatively straight, allowing the use of a circular saw, I always use a Bosch® 1581VS jigsaw to cut just to the line. The saw blows dust off the cutline, its reciprocating blade cuts fast, and its tilting base allows back-cutting. Besides making it easy to trim stock for a tight fit, back-cutting allows the workpiece to be squeezed into place.

I use a block plane and rasps to remove stock up to the cutline, skewing the block plane to reach into dips. Fine-tuning is accomplished with flat and round files.

I've worked with a guy who insists that a belt sander is faster and more controllable than a jigsaw for wasting stock to a wiggly line. Another guy uses an angle grinder. Still another scribes with a small bandsaw, which he outfits with a pair of wheels to make it maneuverable on the job site.

Jim Tolpin is the author of The New Cottage Home *and* The New Family Home, *both published by The Taunton Press.*

Plate Joinery on the Job Site

■ BY KEVIN IRETON

Like most people when they first buy a plate joiner, David Mader, a carpenter in Yellow Springs, Ohio, wanted to find out how strong plate joints really are. Mader crosscut a 2x4 and reassembled it with a pair of no. 20 plates (the largest size available), one over the other. After letting the glue dry, Mader tried to break the 2x4 over his knee. He couldn't do it. Convinced of plate joinery's strength, Mader proceeded to use his plate joiner to butt-join custom flooring that wasn't end-matched.

Often considered the province of shop-bound woodworkers, plate joinery, it turns out, is being used more and more by carpenters on the job site. Plate joinery and biscuit joinery are the same thing, and in this article I'll use the terms interchangeably.

The basic idea behind plate joinery is simple: Plunge a 4-in. circular sawblade into a piece of wood, and you get a crescent-shaped slot. Make a series of these slots along the edges of two boards that you want to join. Insert glue and a football-shaped wooden spline into each slot on one board. Insert more glue into each slot on the other board, then press the two boards together. Water from the glue causes the splines to swell, making a strong, tight joint.

To deal safely with a small piece, this carpenter installed the plinth block first and slotted it in place.

After the slot is cut in the plank, the carpenter taps in a biscuit.

With a matching slot cut in the end of the casing, the carpenter aligns the biscuit and the slot.

Biscuit-Joiner Basics

The typical biscuit joiner is a cylindrical machine, about 10 in. long, and weighs from 6 lb. to 7 lb. It has a D-shaped handle on top and a spring-loaded faceplate in front with an adjustable fence. When the tool is pressed against the workpiece, a 4-in. carbide-tipped blade extends through a slot in the faceplate and scoops out a kerf in the workpiece. You can adjust the distance between the kerf and the face of the workpiece, but any closer than ³⁄₁₆ in., and the biscuit, or plate, may pucker the surface of the wood when it swells. You can also adjust the depth of the kerf to fit the size of biscuit you're using.

Biscuits come in three basic sizes (all three are arcs of the same circle): #0 is about ⅝ in. wide and 1¾ in. long, #10 is ¾ in. wide and 2⅛ in. long, and #20 is 1 in. wide and 2½ in. long. Biscuits are made of beech with the grain oriented diagonally to the length, making them very strong across their width. Biscuits are also compressed so they'll fit easily in the kerf and then swell once the glue hits them. All biscuits are slightly shorter than the kerf they fit into, which not only allows room for excess glue but also provides some play for aligning a joint along its length. This gives biscuit joinery a distinct advantage over doweling as an indexing technique.

Plate joinery works in hardwood, softwood, plywood, particleboard, and even in solid-surface countertops (using Lamello®'s clear plastic C-20 biscuits). Plates can be used in edge-to-edge joints, butt joints, and miter joints.

Joinery Comes to the Job Site

Over time, plate joinery has proven itself strong enough and accurate enough to earn a place in many woodworking shops, where it competes with other joinery methods such as doweling, splining, and mortise-

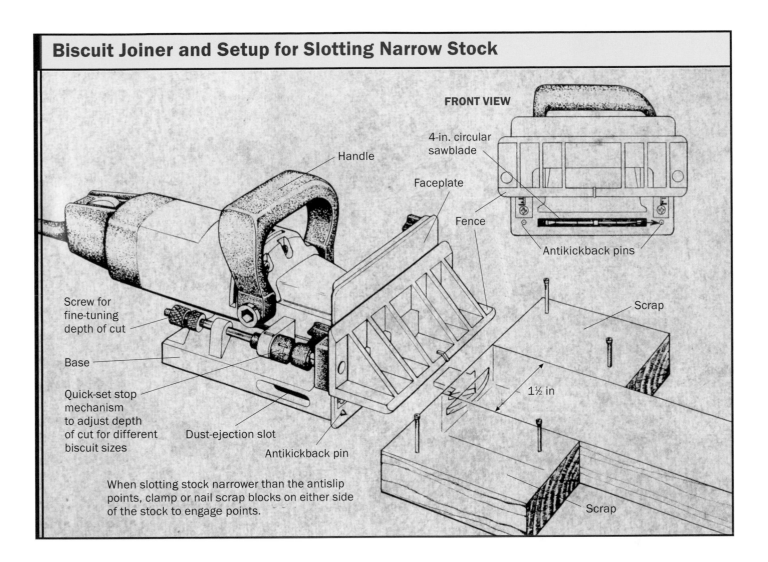

Biscuit Joiner and Setup for Slotting Narrow Stock

FRONT VIEW

Handle

4-in. circular sawblade

Faceplate

Fence

Antikickback pins

Screw for fine-tuning depth of cut

Base

Quick-set stop mechanism to adjust depth of cut for different biscuit sizes

Dust-ejection slot

Antikickback pin

Scrap

1½ in

Scrap

When slotting stock narrower than the antislip points, clamp or nail scrap blocks on either side of the stock to engage points.

and-tenon joinery. The merits of plate joinery relative to these other methods can and have been debated. But most carpenters in the field don't enjoy the luxury of a fully equipped shop, and often their only joinery options are whether to use nails or screws. That's why plate joinery adds a valuable technique to a carpenter's repertoire. After all, a cabinetmaker can successfully argue that a biscuit-joined face frame is not as strong as one joined with mortises and tenons. But no one will argue that adding a biscuit between two pieces of mitered casing (photos, p. 18) won't strengthen the joint or greatly improve its chances of weathering changes in humidity without opening up.

Joint strength isn't the whole story, though. A biscuit joiner is very portable, taking up less room in a toolbox than a

circular saw. And it's extremely fast. Cutting slots and adding biscuits to a mitered door casing might require 30 seconds. Admittedly, even that little time can be significant when multiplied by a houseful of doors and windows. You might consider it worthwhile, though, if you've ever been disappointed when returning to a job to discover gaps in joints that fit perfectly when you nailed them up.

Who's Using Them Where?

Stephen Sewall, a builder in Portland, Maine, feels so strongly about the advantages of biscuit-joined trim that he seldom installs trim without biscuits. On a recent job where he didn't have his biscuit joiner,

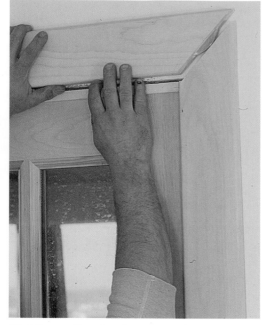

With the tool and the trim registering against the floor, this carpenter makes short work of cutting slots.

By then adding a biscuit spline between two pieces of mitered casing, he prevents the joint from opening as a result of wood shrinkage.

Slightly smaller than its slot, a biscuit allows the carpenter some give for aligning joints.

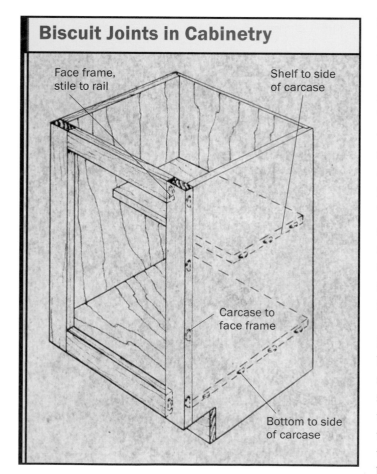

Biscuit Joints in Cabinetry

Face frame, stile to rail

Shelf to side of carcase

Carcase to face frame

Bottom to side of carcase

Sewall nailed up the side casings, but left the head casings loose so that he could add the biscuits later.

Sewall also says that biscuit joinery has made building cabinets on site a lot easier. When installing a fixed shelf in a cabinet, which he used to do by routing dados in the sides to house the ends of the shelf, Sewall can now biscuit-join the shelf to the carcase faster than he can change bits in his router. When biscuit-joining shelves, Sewall often clamps the shelf flat against the upright and registers the biscuit joiner against it as he cuts the slots. Biscuits will also work in ½-in. stock, like the ½-in. Baltic-birch plywood that Sewall uses to make cabinet drawers.

Foster Jones, a partner in Maine Coast Builders of York, Maine, admits that using biscuit joinery adds to the cost of a job and says his company usually decides before starting a project whether to use biscuits. If they do use them, though, they don't just use them on miter joints. They use biscuits to join inside and outside corners of baseboard, to join baseboard to door casing, and to join door casing to plinth blocks (photos, pp. 15 and 16).

When laying a hardwood floor, the carpenters at Maine Coast Builders use biscuits to join the picture-frame border around a fireplace hearth (drawing, below) and to join the border to the flooring that butts into it. They even use the biscuit joiner as a trim saw to trim the bottoms of door casing so that flooring will fit beneath it.

Jones also uses biscuits when fabricating trim for round-top windows. Using biscuits and five-minute epoxy, he joins mitered segments of straight stock end to end in a rough semicircle (bottom right drawing). He usually screws the segments to a piece of plywood rather than clamping them. The five-minute epoxy lets Jones work with the piece after less than an hour of drying time.

Because biscuit joinery relies in part on the biscuits' capacity to absorb water from the glue and swell up to form a tight joint, you may be wondering how well the system works with epoxy, which isn't a water-based glue. Jones and Sewall wondered, too, and broke apart joints that they had assembled with epoxy. Both found the joint to be just as strong as those made with yellow glue, which has a water base. In fact, Bob Jardinico at Colonial Saw, sole U.S. distributor for Lamello joiners, recommends epoxy for

Biscuit-Joining Shelves

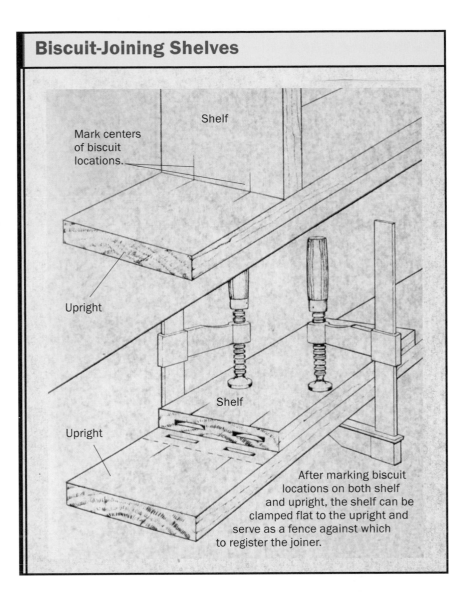

Mark centers of biscuit locations.

Shelf

Upright

Shelf

Upright

After marking biscuit locations on both shelf and upright, the shelf can be clamped flat to the upright and serve as a fence against which to register the joiner.

Flooring Applications for Biscuit Joinery

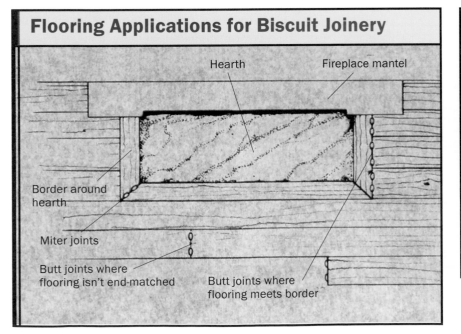

Hearth

Fireplace mantel

Border around hearth

Miter joints

Butt joints where flooring isn't end-matched

Butt joints where flooring meets border

Casing for Round-Top Windows

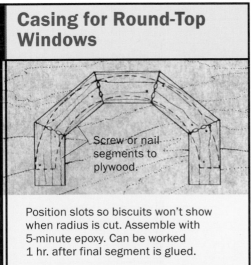

Screw or nail segments to plywood.

Position slots so biscuits won't show when radius is cut. Assemble with 5-minute epoxy. Can be worked 1 hr. after final segment is glued.

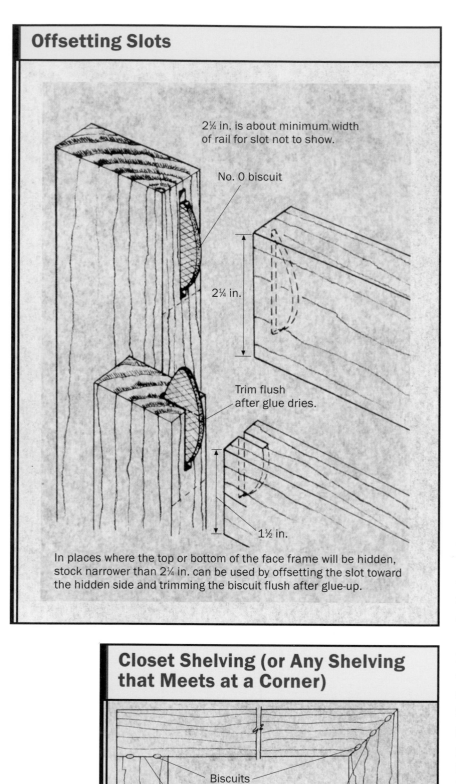

Offsetting Slots

2¼ in. is about minimum width of rail for slot not to show.

No. 0 biscuit

2¼ in.

Trim flush after glue dries.

1½ in.

In places where the top or bottom of the face frame will be hidden, stock narrower than 2¼ in. can be used by offsetting the slot toward the hidden side and trimming the biscuit flush after glue-up.

Closet Shelving (or Any Shelving that Meets at a Corner)

Biscuits

Butt joint

Miter joint

biscuit joinery used outdoors. Makes you wonder if biscuits and epoxy aren't the way to keep mitered handrails on exterior decks from opening up.

Plate-Joining Face Frames

A common complaint when assembling face frames (or cabinet doors) with biscuit joints is that the rail must be wider than the slot for the smallest biscuit. Otherwise the biscuit will show. This limits you to a rail that's at least 2¼ in. wide. Responding to this complaint, Lamello recently introduced face-frame biscuits (H-9) that are ½ in. wide by 1½ in. long. But such a short biscuit means you have to switch to a 3-in. sawblade (also available from Lamello).

A company called Woodhaven makes biscuits out of particleboard that are 1 in. wide by 1⁵⁄₁₆ in. long and for which you cut kerfs with a 6mm slot-cutting router bit. With these same bits you can also use your router to perform conventional plate joinery, but the cutter is exposed and you don't have a faceplate, so you lose some of the safety and convenience of a plate joiner.

As it turns out, though, you can often get away with using a standard biscuit in a narrow rail by offsetting the slot to the outside of the frame and trimming the biscuit flush (drawing, top left). In most cases the exposed kerf and biscuit are either pointing down at the floor and can't be seen, or are covered by a countertop. When cutting slots in narrow stock, it's best to clamp or nail scrap blocks on either side so the steel points on the joiner's faceplate have something to grip (drawing, p. 17). These points keep the tool from slipping during a cut (some joiners employ rubber bumpers or pads rather than steel points).

But Wait, There's More

It's easy to think of other job-site uses for biscuit joinery: mitered ceiling beams, jamb extensions on doors and windows, return nosings on stair treads. You could even use biscuit joinery where two closet shelves meet at a corner and avoid having to screw a cleat to the underside of one shelf to support the other (bottom drawing, facing page).

Do be careful, though, if you decide to buy a biscuit joiner. Beware of the "Law of the Instrument." This is a theory in psychology that states: If you give a small boy a hammer, everything he encounters will need hammering. There may be some things that really don't need to be joined with biscuits.

Kevin Ireton is editor-in-chief of Fine Homebuilding. Photos by the author.

10 Rules for Finish Carpentry

■ BY WILL BEEMER

My first construction job was as a trim carpenter's helper during school summer vacation. My boss had always worked solo, but as he got on in years (I'm older now than he was then), he wanted help moving his tools and materials. All I did that first summer was fetch and carry; I wasn't allowed to measure, cut, or nail. I was told to observe. In doing so, I learned that finish carpentry is essentially a visual exercise.

Finish carpentry makes the eye work hard and skip over imperfections. At this point, the framing carpenter has made the house plumb, level, and square. Or not. A good framer can ease the finish carpenter's job by providing plumb walls and plenty of blocking for nailers for attaching trim. Or not. But even if the framer couldn't read a level and even if he added no more blocking than was absolutely necessary, the finish carpenter's job is to make the doors, windows, and cabinets work, and to make the house look good.

Finish carpentry is more than interior trim. It includes siding, decking, and even roofing—anything the owner will see after moving in. Rough carpenters evolve into finish carpenters by learning how to measure, mark, and cut more accurately. With practice, splitting the pencil line with a saw-cut and working to closer tolerances become second nature.

Perfect miters are only part of finish carpentry. Finish carpenters must develop an eye for proportion and detail. They must learn to visualize the steps that lead to the finished product. I teach these skills to novice carpenters at the Heartwood School in Massachusetts. To make learning them easier, I've organized the following 10 rules of thumb.

1. Avoid Using Numbers

It is usually more accurate to hold a board in place to mark its length rather than to use a tape and involve numbers. Sometimes, using a ruler or tape is unavoidable. I'll use a tape measure on a long piece that's too difficult to mark in place, but generally I don't

Don't measure. It is more accurate to mark trim in place than to measure and then transfer numbers. It's easy to misread a ruler or to confuse numbers while walking to the saw.

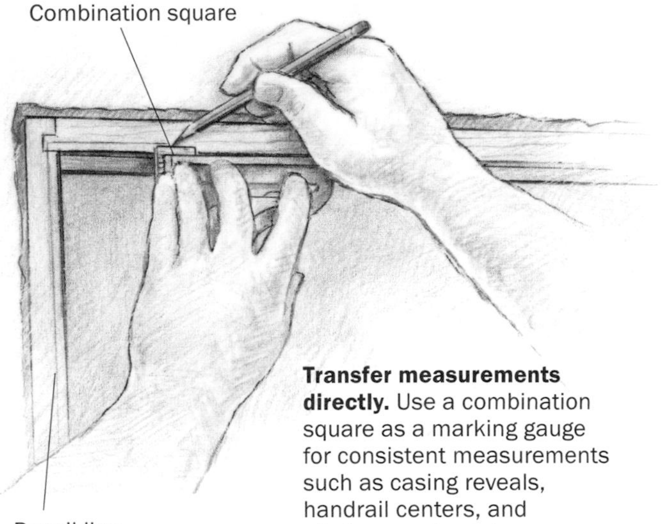

Combination square

Transfer measurements directly. Use a combination square as a marking gauge for consistent measurements such as casing reveals, handrail centers, and window-stool notches.

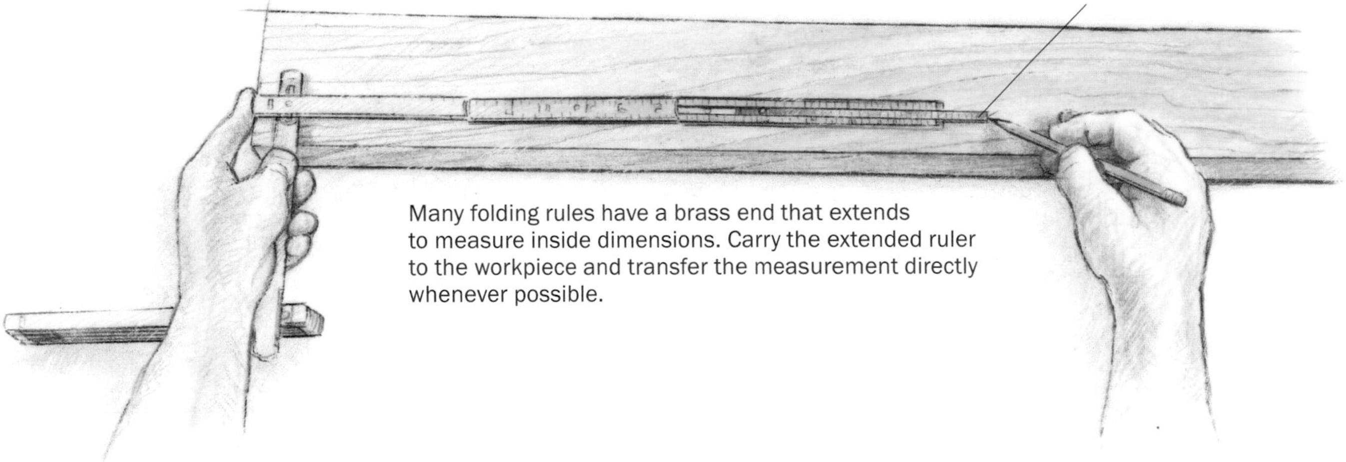

Pencil line indicates where to place the trim.

Brass extension

Many folding rules have a brass end that extends to measure inside dimensions. Carry the extended ruler to the workpiece and transfer the measurement directly whenever possible.

like tapes. A tape can flex and change shape, and the movable end hook bends easily, affecting accuracy.

A rigid rule is better than a tape for measuring lengths less than 6 ft.; hence, the 6-ft. folding wooden rule takes over during trim and cabinet work. The best folding rules come with a sliding brass extension that makes taking inside measurements easy. Open the rule to the greatest length that fits between the points to be measured, and slide out the brass extension the rest of the

distance. Hold it at that length, and carry it to the board to be cut. No need for numbers. Just mark the board from the extended ruler. A combination square or a wood block of known dimension is the best way to lay out the small measurements needed for reveals and other spacings. Learn what dimensions are built in to the tools you use. A carpenter's pencil is ¼ in. thick; you can use it as a spacer for decking. The pencil lead is ⅟₁₆ in. from the edge of the pencil, so it can scribe ⅟₁₆-in. increments. The body of a folding rule is ⅜ in. wide. The blade of the standard combination square is 1 in. wide, and its body is ¾ in. thick.

A door or window should be cased without the use of a tape. Lightly mark the reveal on the jamb with a pencil. Square-cut the bottoms of the casing legs, hold them up to the jamb, and mark the top cuts from the reveal lines. Cut the legs and tack them in place. Miter one end of the head, and holding it upside down over its final position, mark the other end to length.

2. Use Reveals, and Avoid Flush Edges

Wood moves—as it dries out, as the house settles, as you cut it, and as you're nailing it up. It's almost impossible to get flush edges to stay that way. That's why, for example, carpenters usually step casing back from the edge of door and window jambs. Stepping –trim back to form reveals causes shadow-lines and creates different planes that make it harder for the eye to pick up discrepancies.

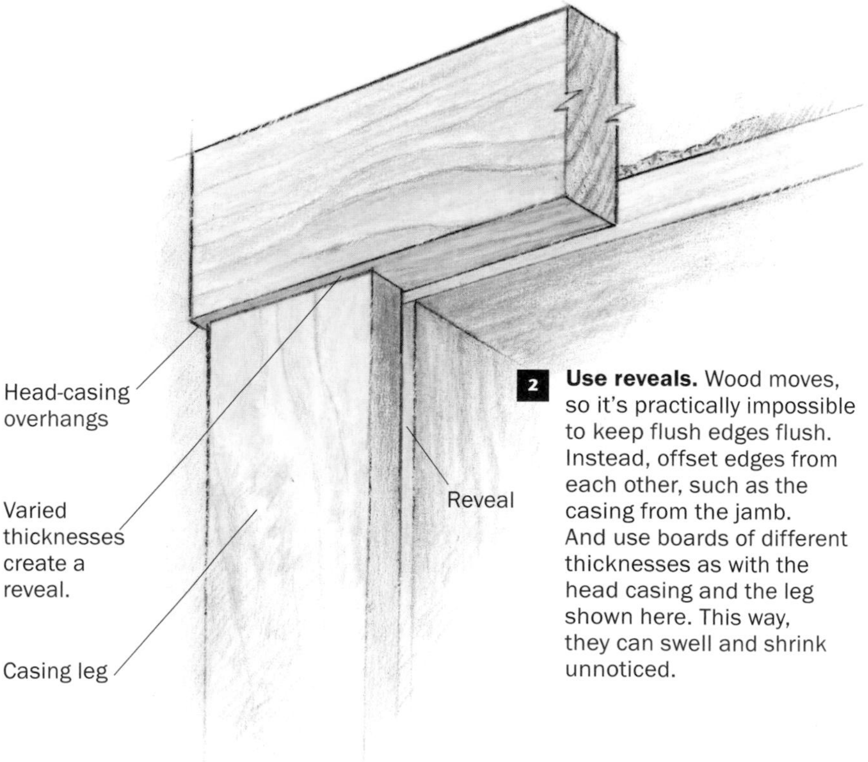

Head-casing overhangs

Varied thicknesses create a reveal.

Casing leg

Reveal

2 **Use reveals.** Wood moves, so it's practically impossible to keep flush edges flush. Instead, offset edges from each other, such as the casing from the jamb. And use boards of different thicknesses as with the head casing and the leg shown here. This way, they can swell and shrink unnoticed.

If a casing is installed flush to the inside of a jamb, it may not stay that way. The eye will easily pick up even a ¹⁄₁₆-in. variation from top to bottom. If the casing is stepped back ¼ in. or ⅜ in., this variation will not be nearly as evident and will be hidden in shadow much of the time. Separate discrepancies, and they become less evident.

In years past carpenters by necessity used trim materials of different thicknesses; planers were not in widespread use. You rarely see mitered casings in older houses because differences in material thickness are obvious in a mitered joint. Instead, the casing legs butt to the head, which runs over and past the legs by ⅜ in. or so. This way, the carpenter didn't worry about the length of the head casing being exact or the side casings noticeably changing width with changes in humidity. The head casing is usually the thicker piece so that the shadow it casts makes it appear to be a cap. Carpenters in the past often placed rosettes at the upper corners and plinth blocks at the bottoms of door jambs. The casings and baseboards butted to them. The variations in thickness of these boards were lost in the overwhelming presence of the thicker plinths and rosettes.

3. Split the Difference

If you're running courses of material between two diverging surfaces, it's obvious that if you start out working parallel to one, you won't be parallel to the other. In the case of decking, roofing, or siding, you can slightly adjust the gap or coverage at each course so that the courses are parallel to the other surface when they reach it.

This adjustment is easily figured. Say that you're shingling an old house, and the roof measures 135 in. from ridge to eave on one end and 138 in. on the other. Divide one of these figures by the ideal exposure per course, 5 in. for normal three-tab shingles. Thus, 135 in. divided by 5 equals 27; this is the number of courses at 5 in. per course.

At the other end of the building, 138 in. divided by 27 yields 5⅛ in. Lay out each side of the roof using the two different increments, and snap chalklines between them. With these adjustments, the chalklines start out parallel to the eaves and end up parallel to the ridge.

It's difficult for the eye to pick up even a ¼-in. change in exposure from one course to another, especially high up on a house. By accounting for discrepancies early and gradually, your adjustment should rarely have to be greater than that.

In cases where the gap or coverage is not adjustable, as in tongue-and-groove flooring, you have to make up part of the discrepancy at the start and the rest at the end. Say you're installing flooring between two walls that are 1 in. out of parallel, and you're leaving a minimum expansion gap of ½ in. between the flooring and the wall. Make the expansion gap 1 in. at each side of the wide end of the room and ½ in. at each side of the narrow end. Shoe molding and baseboards cover the gap (bottom drawing). If you're using a one-piece thin baseboard, you'll have to rip tapered floorboards at the start and finish to keep the expansion gap narrow and parallel to the wall.

In any case, you'll want to use boards as wide as possible as your starting and ending courses. Measure the room width at both the wide and narrow ends, and subtract the expansion gaps. Divide these measurements by the floorboard width. Multiply the remainders by half the board width. These will be the widths of the starting and ending strips at the wide and narrow ends of the room. If these strips are narrow, try adding half a board width. As long as these sums are less than full board widths, use them for the starting and ending strips. This keeps the converging lines of the baseboards and the first and last seams as far apart as possible.

3 **Diverging lines are obvious mistakes.** With shingles or lapped siding, diverging starting and ending points can be hidden a little at a time by slightly tapering the course widths. But this technique doesn't work with other materials, such as tongue-and-groove flooring, whose course can't be easily varied.

Out-of-parallel walls

When installing tongue-and-groove flooring between diverging walls, it is best to plan an equally tapered gap on each side. The angle between the wall and the flooring is halved, making the diverging lines less noticeable.

Start and finish with boards that are as wide as possible. This separates the base and shoe moldings from the first joint between the floorboards, and makes the diverging line harder to see.

Floorboards laid to split the difference

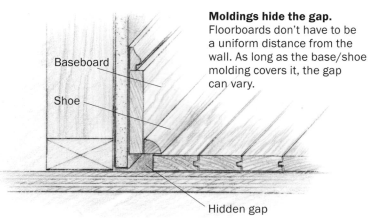

Baseboard

Shoe

Moldings hide the gap. Floorboards don't have to be a uniform distance from the wall. As long as the base/shoe molding covers it, the gap can vary.

Hidden gap

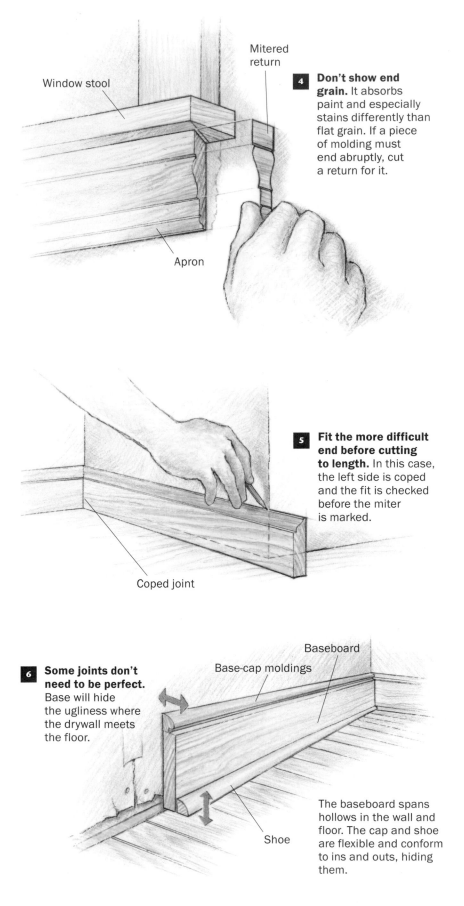

Window stool

Mitered return

4 **Don't show end grain.** It absorbs paint and especially stains differently than flat grain. If a piece of molding must end abruptly, cut a return for it.

Apron

5 **Fit the more difficult end before cutting to length.** In this case, the left side is coped and the fit is checked before the miter is marked.

Coped joint

6 **Some joints don't need to be perfect.** Base will hide the ugliness where the drywall meets the floor.

Baseboard

Base-cap moldings

Shoe

The baseboard spans hollows in the wall and floor. The cap and shoe are flexible and conform to ins and outs, hiding them.

4. Avoid Exposing End Grain

End grain will always absorb stain and paint differently than face or side grain; even if left natural, end grain reflects light differently. Unless you want to emphasize this difference, as with through tenons in furniture, it's best to plan your installation to hide end grain or cut mitered returns to cover it up.

A return is a small piece of trim, often triangular in section, that ends a run of molding. Returns are used in traditional finish work on stair treads, window stools and aprons, butted head casings—anywhere a piece of molding doesn't end in a corner. Small returns are difficult to cut on a power miter saw. The blade often throws the return to some dimly lit, inaccessible corner of the room. I cut them with a small miter box and a backsaw.

5. Fit the Joint before Cutting to Length

If you're coping or mitering a joint on a piece of base, chair rail, or crown, make sure that joint fits well before you cut the other end to length. You may need the extra length if you make a mistake and have to recut the cope or miter. If you had cut the piece to length before miscutting the cope or miter, you'd be grumbling on your way back to the lumberyard instead of calmly recutting the piece.

6. Don't Be Fussy Where You Don't Have to Be

Learn to think ahead to see if what you're working on will later be covered, which is often the function of moldings. At the intersection of wall and floor, for example, the drywall doesn't have to come all the way down to the floor, nor does the flooring need to meet the wall perfectly, because the baseboard will cover the gap. If the floor or wall undulates, you might be tempted to scribe or fill behind the baseboard to follow the contours. In older houses, however, where walls and floors always undulated, you often see three-piece baseboards, with the thin base-cap molding attached to the wall and following its contour while the shoe does the same on the floor. The thicker baseboard installs quickly and easily because it doesn't have to conform; that's what the shoe and cap do.

7. Plan Your Sequence to Avoid Perfect Cuts at Both Ends

There is usually a sequence of installing trim that requires the fewest perfect cuts. For example, with my method of casing doors and windows, only the last cut on the head need be perfect. Cut this end slightly long, and shave it with the chopsaw until it fits just right. One neat trick here: Push the casing

7 **Trimming a room with baseboard and a minimum of perfect cuts.**
By following the numerical sequence in the drawing below, only pieces 2 and 3 require perfect cuts on both ends. The chance of error is reduced by first coping them and then holding them in place to mark their lengths. The copes are planned so that any cracks will be less obvious to people entering the room.

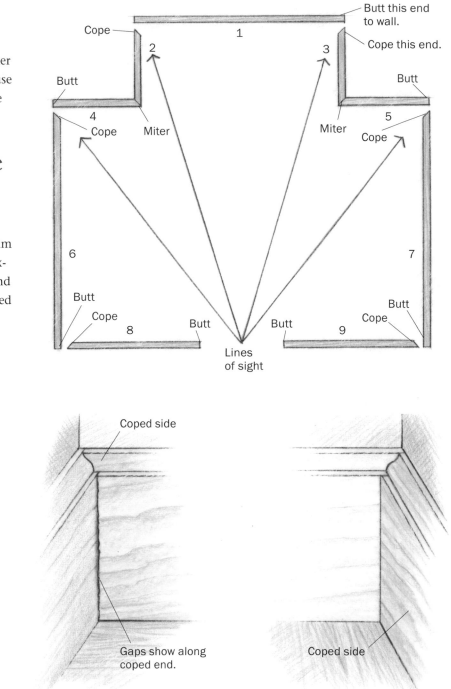

Coped joints. The first piece is butted to the wall. The second piece is mitered as for an inside corner, but the mitered end is cut off where it meets the molding face, leaving a negative of the profile that fits perfectly over the butted piece.

Coped joints look different from different angles. If a coped joint opens up, the crack will be obvious when viewed parallel to the uncoped piece and nearly invisible viewed parallel to the coped piece. Plan the coping sequence so that cracks will be less obvious along likely lines of sight. Cracks will also be less obvious if the uncoped piece is stained or painted before installation; raw wood sticks out.

up to the lowered, idle chopsaw blade. Raise the blade without moving the casing, and then make the cut. The teeth are set slightly wider than the body of the blade, so the cut will take off ¹⁄₃₂ in. If you had installed the head first, you then would have had to make an exact miter cut on each casing leg to make the joint turn out right.

As with trimming doors and windows, sequence of installation is also important when running trim around a room, whether it's baseboard, chair rail, or crown molding. For instance, I'm right-handed and can generally do a faster, neater job of coping the right end of a board than the left end (coping is an alternative to mitering joints at inside corners; bottom drawings, p. 27). Consequently, I often prefer to work from right to left around a room.

As I work my way around a room, especially when running crown molding, I'll often end up with a piece that needs to be coped on both ends, a challenge for even the best carpenters. I try to plan my installation so that this last piece of trim is in the least conspicuous place. If a coped joint isn't perfect, or if it opens up over time, the crack is most visible when viewed at right angles

to the coped piece. Wherever possible, I orient the coped pieces so that people entering or using the room won't have right-angle views of them.

8. Parallel Is More Important than Level or Square

Some rules of carpentry change from framing to finish work. Instead of keeping track of plumb, level, and square, you now must keep finish materials parallel to the walls and floors. The eye sees diverging lines more readily than it sees plumb and level.

The only exceptions are cabinets and doors, which must hang plumb to work properly. If the floor isn't level, trim the door bottoms parallel to the floor rather than leave them level with a tapered gap.

If a deck is out of square, run the decking parallel to the house wall rather than squaring it up to the deck framing. If for some reason two lines must diverge, separate them as widely as possible so that the difference is harder to see.

8 **But I hung it plumb.** A level door bottom over an out-of-level floor has an eye-catching, tapered gap at the bottom. Trim the door bottom so that it's parallel to the floor. It won't be level, but it'll look good.

Scribe the door bottom parallel to the floor.

Unlevel floor

Trimmed parallel to the floor, the door looks right.

9 **Nothing is random.** Even something as simple as decking benefits from thoughtful layout. The randomness (left) looks sloppy compared with careful layout (right).

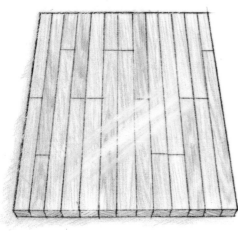 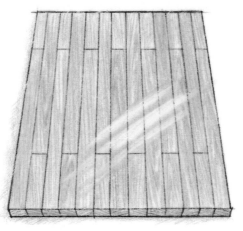

9. Nothing Is Random

Whenever I find myself saying, "It doesn't matter," the red flag goes up. Which end of the board you cut first, which face is out, where you put the nails—all this matters, and the care you put into the details shows up as craftsmanship in the entire job. "God lives in the details," said Mies van der Rohe, and this is especially true in finish carpentry. Occasionally, it won't matter, but you should first consider whether it does. Taking this to the extreme may result in a phenomenon I call "analysis paralysis." As your experience increases and your eye becomes more efficient, it will become second nature to line up your nails in an attractive pattern and to look critically at each board as you carry it to the saw.

10. Finish the Job

A contractor usually has to complete a punch list before final payment is issued, but sometimes getting all the details wrapped up is like pulling teeth. The clean-slate attraction of starting a new job can overpower the drudgery of completing the old. This temptation can backfire, souring good clients and losing referrals.

Owner-builders doing their own work are often tempted to move in to their house before the finish work is done, thinking it will be easier to do when it's close at hand. After a while, they don't notice the lack of trim, especially if there's furniture in the way, and it becomes harder and messier to set up the tools and work around the obstacles. It can be a strain on a marriage if the bathroom doors aren't hung after a few years of residence. I advise owner-builders to get everything done before they move in, and contractors to finish all work before they move on. They'll be glad they did.

Will Beemer is executive director of the Heartwood School in Washington, Massachusetts, and executive director of the Timber Framers' Guild of North America.

Pneumatic Finish Nailing

■ BY JIM BRITTON

Three decades of pneumatic finish nailing have come and gone. Little that's fundamental has changed in the mechanics, use, and care of these tools. The greatest change is that carpenters will never return to hand-nailing. Hand-nailing is too time-consuming, particularly for trim carpenters who also must set the nails. Pneumatic nailing is cleaner. The marring of finished surfaces is almost nil.

Every finish carpenter I know now has at least one pneumatic nailer. In a typical day, I use two or three nailers for tasks that range from hanging doors to nailing base quickly to tacking on small molding returns.

Pneumatic Nailers Deserve the Same Respect as Firearms

Pneumatic nailers will fire a nail 2 in. into a piece of wood—or into your body—and return to the ready position in 125 milliseconds. They rate the same respect that you would give a gun.

Never point the nailer at others. Always assume that the tool is loaded and ready to shoot. Safety glasses are clearly a priority, not just to deflect an errant nail but also to protect your eyes from the airborne dust and debris kicked up by the discharge of air from the exhaust port of the tool. Corrective eyeglasses also can serve as safety glasses, as long as they are either tempered glass or polycarbonate. Better safety glasses, though, have side shields to protect your eyes from projectiles coming from the side.

As a safety feature, almost all the nailers on the market have a contact element at the discharge point. This contact element must be depressed at the time you pull the trigger to fire the nailer. I have seen some carpenters disable this contact element, presumably to speed up nailing or perhaps to reduce the slight marring of the wood that an unpadded contact element can produce. No matter what the reason, disabling the contact element is just too dangerous. Without a contact element, triggers on pneumatic nailers are unguarded, and an errant touch can shoot a nail unexpectedly.

Clean, Dry Air and Regular Lubrication Are Key

These nailers will last nearly forever if properly cared for. Pneumatics like clean air. Dust in the airstream is abrasive, and it speeds up the normal wear and tear of a nailer's moving parts. I clean my compressor's air-intake filter regularly, following the compressor manufacturer's procedures. I tend to do it more often in dusty environments.

Pneumatics also like dry compressed air, although the latter phrase is a bit of a misnomer. Limiting moisture buildup is the real goal. As incoming air is compressed, moisture is literally squeezed out and condenses. Most of this water collects in the compressor's tank, but enough can make it into the hose to cause nailer trouble.

Limiting moisture buildup is mainly a matter of draining the compressor of condensed water daily. I do this chore by running the compressor with its tank drain open for a few minutes every morning. This task is especially critical when the weather turns

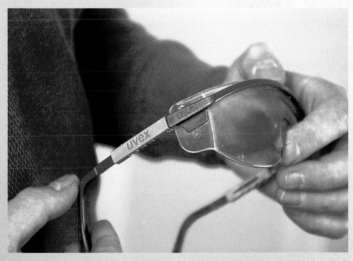

Protect your eyes. Pneumatic nailing is practically a necessity to be competitive. The dangers, however, particularly to the eyes, are real. Shatterproof glasses, preferably with side shields, are a necessity.

Pneumatic nailing has replaced hand nailing for trim carpenters for nearly everything from hanging doors to installing moldings.

How Pneumatic Nailers Work

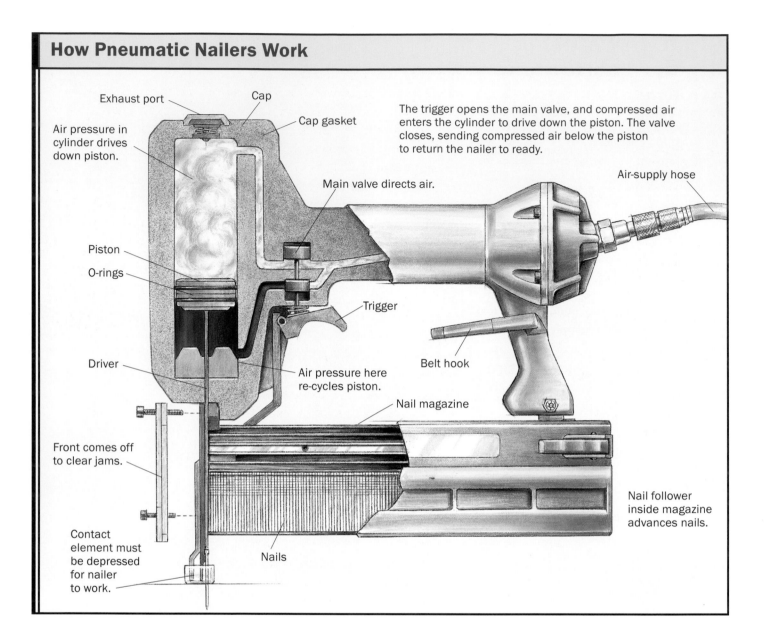

Exhaust port

Cap

Cap gasket

Air pressure in cylinder drives down piston.

The trigger opens the main valve, and compressed air enters the cylinder to drive down the piston. The valve closes, sending compressed air below the piston to return the nailer to ready.

Air-supply hose

Main valve directs air.

Piston

O-rings

Trigger

Driver

Belt hook

Air pressure here re-cycles piston.

Nail magazine

Front comes off to clear jams.

Nail follower inside magazine advances nails.

Contact element must be depressed for nailer to work.

Nails

cold because water in the lines or nailers can turn to ice and block the flow of air.

If you must use your pneumatics in arctic weather, a Paslode® spokesman recommends a tablespoon or so of automotive antifreeze be added to the compressor tank. Antifreeze mixes with any water vapor in the compressed air, preventing the tool from freezing up. I'm careful with antifreeze; it is toxic. And I'm sure not to use an antifreeze that contains antileak compounds. They can clog small ports in the nailer.

Even if you keep water, and therefore ice, out of your nailers, cold weather can make the rubber O-rings that seal nailers less flexi-

ble. To avoid damaging the O-rings and to ensure trouble-free operation, I warm my nailers before work. Bringing them inside overnight is a good idea, as is keeping them near the truck's heater for the ride to work. I warm the nailers during lunch, too.

I keep an air nozzle at hand and use it to dust my nailers daily. Owning nailers with both steel and plastic magazines, I think the plastic type needs more cleaning.

Air leaks anywhere in the system reduce nailer performance and run the compressor more, increasing wear and tear. I'm sure to keep all caps and screws tight on my nailers.

Audible leaks are almost certainly reducing nailer performance, so I fix them promptly.

For light use—when, for example, I'm cutting trim and nailing it up as I go—I lubricate nailers daily with three drops of pneumatic oil through the air intake. On a heavy-use day, when I'm constantly nailing precut trim, I lube my nailer again after lunch. Overlubrication will yield an oily exhaust port. Too little oil, and you may smell the O-rings overheating and wearing out. With experience, you should be able to dial in the correct amount of oil.

Automatic oilers that install on the nailer are available. However, they are automatic only if they have oil in them. They usually hold enough oil to supply one nailer for several weeks. I think that daily oiling is an easier habit to get into than is, say, biweekly oiling. Automatic oilers carry with them a real risk of complacency.

Some manufacturers make oilless nailers that have permanently lubricated seals and guides. Never add oil to them because oil can dissolve the nailer's seals.

Even Correctly Maintained, Nailers Occasionally Break

The simplest breakdown is a jam. Jamming is usually caused by nailing into a drywall screw, framing clip, or some other hidden fastener. The nail curls inside the tip of the nailer, stopping the driver from cycling and the other nails from advancing. To remove a jam safely, I first disconnect the air supply. Most nailers have a clip that opens the front of the tool. With this clip open, I remove the jam, close the tool, and reconnect the air hose.

Sometimes nails stop advancing in the magazine. This problem usually means the magazine is dirty, and the fix is cleaning it with WD-40®. If the magazine is clean and nails still don't advance, the problem is probably a worn magazine or nail follower, or you're using the wrong nails.

Nailer breakdowns that require disassembling the tool for repair are usually one of two types: The driver breaks, or a seal breaks, causing massive air leakage. Either problem can usually be fixed on the job if the parts are available. Every carpenter ought to carry an O-ring kit and spare driver for each nailer. These kits can cost from $20 to $100, depending on the manufacturer. Still, they're cheap insurance compared with driving around, shopping for parts instead of working.

If your nailer's not leaking and won't fire, the piston may be stuck at the bottom of its stroke. First, try re-cycling the piston by letting the compressor run up to full pressure with the air hose disconnected from the tool. Next, hold the nailer on a nailing surface with the contact element depressed and the trigger pulled, and reconnect the air. This step will send a strong surge of air to the underside of the piston, returning a stuck piston to its normal position.

If the nailer stops setting nails yet the nails in the magazine seem to move, the driver might be broken. Air will not be leaking, and you may see the tip of the driver sticking out.

If you see the tip of the driver and re-cycling doesn't retract it, disconnect the air and grab the tip with pliers. If the tip comes out, it's new-driver time.

If the nailer isn't setting nails fully, that could indicate that the nails are too long, the air pressure is too low, or the driver is worn and needs replacement.

Large air leaks reduce pressure over the piston and yield poor performance. These leaks can happen anywhere, and you can often hear where the air is leaking. Leaks at the trigger are fairly common. Fixing a leak from the cap of the tool can be as simple as tightening the cap screws, but cap gaskets can fail and require replacement. When you have an idea where the leak is, take the tool apart and inspect for worn or torn O-rings. I always lube replacement O-rings with Lubriplate® GR132 grease (available at Paslode dealers) as they're installed.

Pneumatic nailers will fire a nail 2 in. into a piece of wood—or into your body—and return to the ready position in 125 milliseconds. They rate the same respect that you would give a gun.

Three drops of oil a day is usually enough. Use too much, and you'll notice an oil buildup at the exhaust port. Use too little, and you may smell the rubber O-rings wearing out. Proper lubrication takes a little practice.

If you can pull out the driver, it's broken. The fix for this problem requires the nailer to be taken apart. Most nailers are simple enough that stripping and reassembling can be done on the job.

Nail jams are the most common glitch. Caused frequently by nailing into hidden metal, jams are easily cleared from most nailers by pulling back the nail follower, unhooking the air, and removing the nosepiece.

There is no AAA for nailers, so you've got to carry spare parts. O-ring kits are available from most manufacturers. Keeping one in your trunk enables job-site repairs that can save hours of downtime.

Choosing the Right Nails

Inherent in the smooth operation of pneumatic tools is proper nail or staple selection. There is an ideal fastener for every application. The simplest consideration is choosing a nail that's just long enough for the job (photo, right). I generally use nails long enough to penetrate the trim and drywall, then extend ¾ in. into the underlying framing. Unnecessary nail penetration uses more air and more steel. You pay for both.

Generally speaking, the harder the wood, the shorter the nail should be. For example, ½-in. penetration into hardwood substrate is usually adequate. Harder woods hold nails better, so there is less need for deep penetration. The only exception to this rule occurs when nailing a hardwood to a softwood. Also, driving too long a nail into hardwood is difficult, and the nailer might not have enough power to set a nail in such material.

Shiners are nails that shoot from the finished face of the trim or board. They happen most frequently in situations when you're nailing into the edge of a board (for example, when nailing through casing into a door jamb). Sometimes they happen because you've aimed the nailer incorrectly, but that's rare with experience. Most often, they occur when hard grain deflects the nail's point. This problem is avoidable with experience.

Most nails have a chisel point (photo, right). Correctly orienting the nailer to the wood grain can nearly eliminate shiners and reduce splitting. To orient a nailer correctly, examine the nails, and position the nailer so that the long axis of the nail's chisel point enters the wood perpendicular to the grain. This way, the point of the nail cuts or punches through the wood, rather than being deflected by it or splitting it.

Use the correct nail length. For most trim applications, the author uses nails that penetrate ¾ in. into the framing. Less won't hold, and more is hard on the nailer.

TIP

Keep an air nozzle handy and use it to clean your nailer. It's quick and prevents dust buildup.

Like their namesake, chisel-point nails are driven parallel to the grain split wood. The solution is to orient the nailer to drive these nails perpendicular to the grain direction.

Nails at stress points add strength where it's most needed. Here, the author nails in a spot that will be hidden by hardware, saving the painter some effort. Near hinges is another spot that requires nailing.

Follow the path to faster nailing. When nailing moldings such as this casing, the author runs his nailer along a linear detail to locate nails quickly and accurately.

My Favorite Techniques

I'm careful to make my nail pattern on one side of a door or trimmed opening mirror that of the reverse side. I think that patterned nailing holds material evenly, and because missing a nail breaks the pattern, it's easier to spot a place where I forgot to shoot a nail. I use pairs of nails on casing, base, door jambs, and crown molding. Door jambs and casings are nailed near stress points, such as hinges and strike plates (photo, top left).

When nailing a molding that has a pronounced crease or edge detail, such as a colonial-style casing or crown, I use the edge detail to guide my nailer (photo, top right). I depress the contact element of the nailer to firing position and slide it along while nailing, never lifting. This way, I just need to pull the trigger when the nail spacing is correct. I don't need to stop moving the nailer.

Find one stud that's on layout, and you can rock and roll. By placing the end of his tape in line with one stud, the author need only align his nailer with the tape's layout marks to know the nails are hitting studs and securely anchoring the base.

The Nail You Need Determines the Nailer You Need

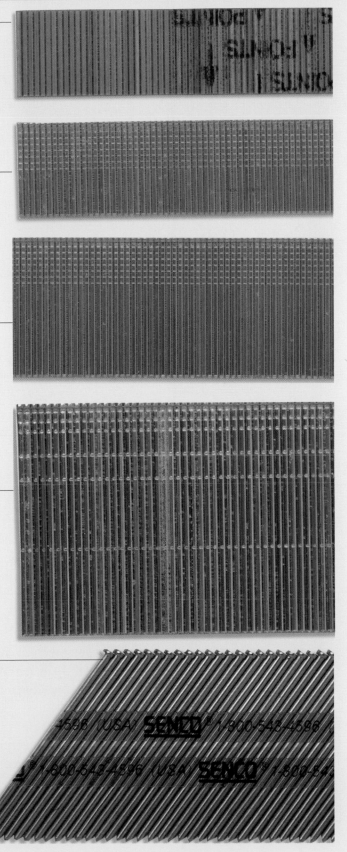

Pin nailers (18 ga.)

Headless pinners are great when fastening tiny trim pieces. With no head, the pin is virtually invisible. Puttying nail holes is unnecessary. There is little resistance to pull through. Use for wood-to-wood connections; pin lengths are from ½ in. to ¾ in.

Slight-head pinners (18 ga.)

Slight-head pins resist pull through better than headless pins. The head is so small that putty is seldom used. I like my slight pinner for stain-grade toe-kick and end-panel veneers during cabinet installation. Use for wood-to-wood connections; pin lengths are from ⅝ in. to 1 in.

Brad nailers (18 ga.)

The brad nailer's headed nails make it more versatile than either pin nailer. The drawback is the larger puncture to be filled. These nailers can shoot through drywall into framing. However, I recommend these small nails only for low-stress trim boards. Brad lengths are from ⅝ in. to 1⁹⁄₁₆ in.

16-ga. nailers

The 16-ga. nailers are great for the one-nailer carpenter. These tools will perform nearly all interior-trim tasks. Because their rectangular cross section increases the contact area, I believe 16-ga. nails equal the holding power (but not the shear strength) of round 15-ga. nails. Nail lengths are from 1 in. to 2½ in.

15-ga. nailers

The 15-ga. nailers are the most common finish nailer on the job site. Most are the angled type because this shape allows these bulky nailers into tight spots. Their large-gauge nails offer excellent structural integrity, making these nailers my first choice for solid-core door and closet-cleat installation. They can be used for most interior trim. Nail lengths are from 1¼ in. to 2½ in.

When nailing down a molding that has a pronounced crease or edge detail, such as a colonial-style casing or crown, use the edge detail to guide the nailer.

Fastening Molding to Drywall

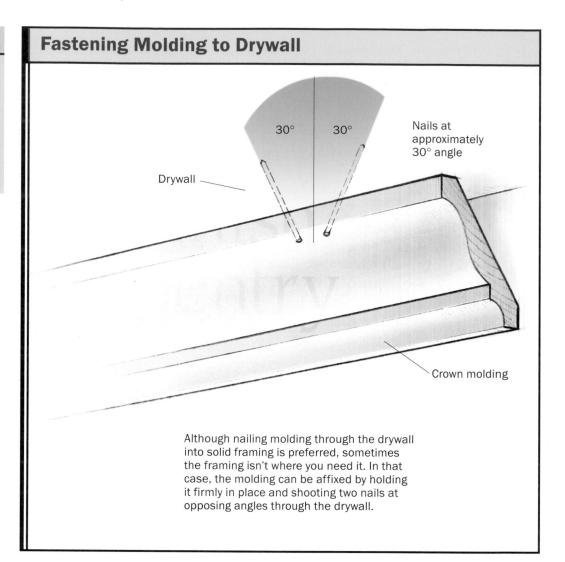

30° 30°

Nails at approximately 30° angle

Drywall

Crown molding

Although nailing molding through the drywall into solid framing is preferred, sometimes the framing isn't where you need it. In that case, the molding can be affixed by holding it firmly in place and shooting two nails at opposing angles through the drywall.

When nailing tapered casing such as ranch or colonial style, I use shorter nails, usually 1¼ in., to attach casing to jamb than to attach casing to the wall. There, I use 2-in. nails.

When nailing base, I locate one stud in a wall that I'm sure is on a regular layout. I lay my tape on the floor with its end by that stud. Then I nail the base to the wall studs, now located by the layout marks on my tape (bottom photo, p. 36).

When it's available, I always nail into a framing member. But it isn't always there. A classic situation is crown-molding installation. Few framing carpenters supply backing where it's most needed for crown: in the ceiling at walls parallel to ceiling joists.

When this situation arises, I push the crown tight to the ceiling and shoot two nails into the drywall in opposite directions, at perhaps a 30° angle and about ½ in. apart (drawing, above). I take care to keep the nails from converging in the same area of drywall. The bottom of the crown will be nailed into either the wall top plates or the studs.

Clamping prior to nailing will help to align crucial joints and improves any glue bond. Pneumatic nailers aren't great at pulling together joints, but they will usually hold together one that has been clamped (top photo, facing page).

In a related situation, a mass (such as a hammer) is often helpful when nailing

Try These Nails

Buying quality fasteners will reduce jams and misfires. The surest bet is to use those made by the manufacturer of your nailer. However, there are many manufacturers of nails, and many work well in different makes of nailer. Below are the brands that I've found can be interchanged successfully.

- 18-ga. pins, 0.049 in. dia.: Senco®, Duo-Fast®, Fasco™, Stan-Tech
- 18-ga. brads, 0.049 in. dia.: Senco, Duo-Fast, Airy®, Fasco, Stanley®-Bostitch®
- 16-ga. nails, 0.059 in. dia.: Paslode, Duo-Fast, Prebina, Hilti®, Fasco, Stanley-Bostitch
- 15-ga., 33° angled nails, 0.069 in. to 0.072 in. dia.: Senco, Fasco, All-Specs, Porter-Cable, Airy, Interchange
- 14-ga. nails, 0.077 in. to 0.080 in. dia.: Senco works in most 14-ga. nailers

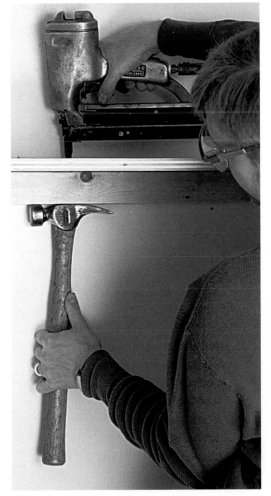

Pull the work together, then nail it. Pneumatic nailers don't pull pieces together; in fact, they're likely to push them apart. With pneumatic nailing, it is useful to hold the workpieces together with clamps or to back up the assembly with a weighty object such as a hammer as the pieces are being nailed.

wood-to-wood connections (photo, right). Backing up the bottom component with something heavy will keep it from jumping away from the top piece when the nail is fired.

Finally, I've discovered that as I switch among nailers, changing line pressure is not necessary. I leave the regulator set at the highest pressure I need to use that day. Even though pinners and bradders need only 70 psi to function well, they also work fine at the slightly higher pressures used for the larger nailers.

Jim Britton is a carpenter and trim contractor in Jacksonville, Oregon.

A Pair of Built-In Hutches

■ BY KEVIN LUDDY

One thing I've learned from working as a carpenter for 20 years is that homeowners are uncharted waters. When I met with the owners of this house about building a pair of hutches, two things became apparent. First, even though they wanted to keep the price down, they wanted a quality job. This was good.

But the second thing was that these hutches had to fit into two alcoves that flanked an existing fireplace, and these alcoves were ridiculously out of square. And although they were nominally the same size, the two spaces were more than 2 in. different in both height and width, a fact that threatened thing number one. Another threat to the economics of the project was that the widths of "about 4 ft." (as I had been told on the phone) were actually 49 in. and 52 in., not great for efficient use of sheet goods.

A Dry Run in the Shop

Each of the paint-grade hutches would consist of a base section with cabinet doors and a countertop, and an upper section with open, adjustable shelves. The cases and

shelves would be ¾-in. birch veneer-core plywood, and the doors and moldings would be solid wood. To save material, I planned to use the existing walls in the two alcoves as the backs of the hutches.

I prefabricated all the parts for the hutches in my shop. I also made the raised-panel doors for the bases and the trim pieces that would hide the variations in the walls and ceiling. I preassembled the bases and shelf units in the shop to make sure everything would go together easily, and then the parts were sanded and given a good coat of primer.

Lots of Shims and Scribing

When I arrived at the site, a finished house, the first things I unpacked were drop cloths. I reacquainted myself with the site dimensions and peculiarities and then located the framing behind the drywall for attaching the case sides.

With the existing walls acting as the backs of the cases, the first pieces I installed were the sides of the lower cabinets. At the shop, I'd notched each side for the toekick. The floor in one of the alcoves was fairly

Faced with spaces that weren't close to square, this carpenter judiciously applied trim and careful scribing to hide the worst of it.

The wall in one of the alcoves was out of plumb almost ⅜ in. over the height of the lower cabinets.

After the sides of the lower cabinets had been fit to the back wall, the bottom shelf was glued and nailed to cleats on the sides.

With the existing drywall forming the back wall of the new hutch, the sides with the bottom shelf attached are centered in the opening and shimmed plumb.

After the lower-cabinet sides and shelves are put in, a 1x cleat is installed to support the back edge of the countertop.

level, so I plumbed the sides in place and scribed them to the back wall, keeping the top edges level (more on scribing later). When I was satisfied with the fit, I glued and nailed the bottom shelf to the two sides (top photo, facing page). Then I centered the assembly in the opening, shimmed each side plumb and shot nails to hold each side in place (bottom photo, facing page).

The floor of the other alcove was not level, so before scribing the back edges, I had to shim the lower of the two sides up from the floor. The height differences were to be hidden by a custom-fit toekick installed later. Next, I screwed a cleat to the back wall on each side to support the back edge of the countertops (photo, above).

When fitting the countertops, I slightly undercut the back 12 in. on both sides of each top (which would be hidden under the upper cases) to allow for easier fitting. The tops were then scribe-fitted to the back wall and nailed in place. Before going any further, I glued and nailed the rails under the countertops to give them added strength during construction (photo, right).

Before continuing with the construction, the author installs rails under the front edges of the countertops for added strength.

For most scribes, I prefer a small electric grinder because it is easy to control and to feather to the line.

Using a straightedge and a level, the author tacks shims to the wall to create a plumb line before the sides are fit.

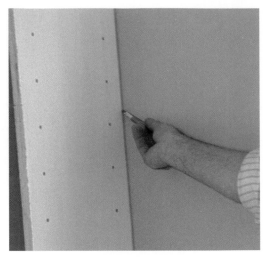

To scribe variations of less than ⅛ in., the author lets a pencil ride on the wall, its point recording the scribe line.

An electric grinder creates a bit of dust, but the grinder is easy to control and allows a feather edge to the scribe line.

Fitting the Top Pieces Is a Grind

Before fitting the sides for the upper-shelf sections, I tacked shims to the drywall, using a level against a straightedge to keep the shims in a plumb line (photo, left). The sides were the toughest pieces to fit because they were captured on three sides by the ceiling, the back wall, and the countertop, and all these joints were visible. Starting with the tallest side (so that I could use it in a shorter spot if I screwed up), I scribed each side piece to the textured ceiling first.

Then I set each piece in place at a slight angle because of the tight top-to-bottom fit and pencil-scribed each piece to the back

To make the sides perfectly parallel, a shelf is set in place, and the shims are adjusted to the shelf.

wall (top right photo, facing page). For this scribe of less than ⅛ in., I held a short pencil to the wall and let the point ride on the piece to be fit.

I've seen many different tools for cutting scribe lines: jigsaws, handplanes, rasps. For most scribes, though, I prefer a small electric grinder (bottom right photo, facing page). It may create a little more dust, but I think a grinder is easier to control and to feather to the line.

The side pieces were then tacked in place. To keep them parallel while they were being nailed, I rested a shelf on its supports between the sides and shimmed as needed (photo, above).

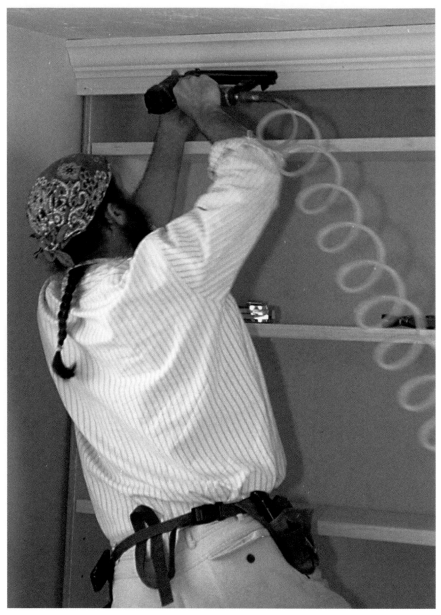

The first trim to be installed is the crown backer and molding, which is adjusted to disguise a ceiling that is out of level.

Hiding a Multitude of Shims

I'd now reached the magic portion of the show, installing the molding and face frames to finish the piece and to camouflage all the differences in height and width. I started at the top with the crown molding and the backer board installed underneath.

Much to my amazement and luck, the ceiling above the left-hand hutch was level (all the error was in the floor). So the crown backer and molding went in quickly and easily. The installation gave me a reveal of

about 1¾ in. of backer board below the crown. My luck ran short on the right side, where I needed to make up almost ½ in. in 4 ft. I decided to hide this difference in degrees (photo, left). First, I applied the backer board ¾₆ in. out of level and then nailed the bottom edge of the crown, varying the reveal from 1¾ in. to 1⅞ in. To hide the last ⅛ in. or so, I twisted the crown into position, making it slightly taller on one side than the other. When I stepped back from the hutch, I was impressed by how even the detail looked after all that fussing.

My bad luck continued for the upper face-frame stiles. I thought a 1⅜-in. width would give me plenty of stock to scribe to the wall and extend beyond the shelf sides by at least ⅜ in. But because I had to shim the ¾-in.-thick sides off the wall by as much as ½ in., the face frames barely covered the sides. I would have to make new face frames.

At my urging, the homeowners had decided on fluted stiles for the lower cabinets. The walls on the left side had little taper, so the fluted stiles went in with little tweaking. I biscuited the tops of the stiles into the rail and then glued and nailed them to the case sides and bottom.

I had made wider stiles for the right-hand cabinet, and I needed every bit to make up for more than ⅜-in. taper in the walls. Scribing these stiles took a bit more care so that I would be left with the exact finished opening for the doors when I was done. I made a couple of test fits and then biscuited, glued, and nailed the stiles in place. After fitting the doors, I called it a day.

Face Frames Stabilize the Shelf Sides

The next day, I returned with the wider upper face-frame stiles. Increasing their width to 1¾ in. allowed me to scribe them to the wall with plenty of extra to hide the ends of the shelves. To scribe a taper of this magnitude, I grabbed my 99¢ compass-style

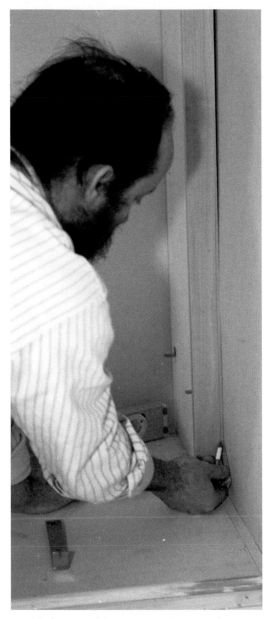

For big irregularities such as the ones here, inexpensive but invaluable compass-style dividers transfer the wall's contour to the piece being fitted.

To help stabilize the sides of the shelf sections, the face-frame edges are coated with construction adhesive before being nailed to the shelf sides.

dividers (photo, left). I glued and nailed the stiles in place and put construction adhesive on the wall edge of the stiles to help stabilize the sides (photo, right). I then gave the doors a final adjustment, filled the nail holes, and touched up the primer.

After the painter put on the finishing touches, I got a call from the clients saying that they loved the way the hutches turned out. The scribing had hidden the worst of the variations, and the trim details pull the eye away from the rest. The homeowners had picked out knobs for the doors, and I promised to put them in next time I was in the neighborhood.

Kevin Luddy runs Keltic Woodworking, an architectural woodworking company in Wellfleet, Massachusetts.

Curved Baseboard Corners

■ BY ERIC BLOMBERG

Thin-coat plaster is increasingly popular for interior walls in northern California where I work, but the variable textures and the bullnose corners of these walls can raise havoc when it's time to apply trim. I discovered this a couple of years ago when the crew I was working on started to install baseboard in a house near Santa Rosa, California. Each outside corner on these coarse-textured walls had a ¾-in. radius. Mitering the baseboard around these corners was out of the question. Square corners would look bad juxtaposed with the bullnose walls, and mitered corners would leave an awkward gap between the wall and the baseboard. Pondering paradigms led us to our current solution.

Our technique requires only a biscuit joiner, a simple jig, and rounded corner stock that we have run off at a local mill shop. The back of the rounded corner pieces are concave, so they fit snugly against the radius of the plaster walls at the corner. That eliminates the triangular gap that would result if the baseboard were mitered at 45°. In this house we butted the 1x6 maple base-

board into the rounded corner pieces, so the baseboard follows the wall cleanly, even if the corner isn't exactly 90°.

It took trial and error to get the technique right. We realized that rounded corner stock was part of the answer; the trick was learning how to join the baseboard and the corner pieces together cleanly. We resolved the problem with biscuit joinery, using a jig we devised. The technique is fast and effective.

Corner Pieces from a Mill Shop

The corner pieces were created with a single pass through a multiple-head molder, producing 10-ft- and 12-ft.-long pieces that we cut to length on the job. Because the shop grinds knives for each job, it could easily make radiused corner pieces. You can also make the pieces yourself (sidebar, p. 50).

For baseboard that will be painted, grain direction in the corner pieces is irrelevant. That was the case on this job. But the mill shop we use also can produce corner pieces

A Jig for Joining Rounded Baseboard Corners

Fence set at 90°

Biscuit joiner

¾-in. plywood

Corner piece

Rounded corner pieces help baseboard fit plaster walls with bullnosed corners. Corner pieces are biscuited to straight runs of baseboard, and a simple jig ensures a perfect fit. The jig, made from ¾-in. plywood scrap, holds a rounded corner piece in position so that the biscuit joiner can plunge the cuts for two #10 biscuits. The fence on the biscuit joiner should be set at 90°. The thickness of the scrap used to make the jig should equal the radius of the corner piece.

with the grain running horizontally, just like the base. That would be helpful if the trim were going to get stained, and the grain direction had to match.

Corner Pieces First

We have a three-step procedure for fitting baseboard, and we start with the corner pieces. After cutting the pieces to match the height of the baseboard (drawing, above), we put them in our jig and mark the locations for two #10 biscuits (biscuits come in several sizes, ranging from #0, the smallest, to #20, the largest). Using #10 biscuits with 1x6 baseboard means we can get two biscuits at each joint, and the slots aren't deep enough to break through the face of the corner pieces. (We cut the biscuit slots in the baseboard later, after all of the pieces have been cut to length.)

The trick to cutting the slots accurately is the jig; it holds the corner piece and aligns the cutter in the biscuit joiner. The jig is simply two pieces of ¾-in. scrap plywood screwed to a base. The corner piece fits between the two scrap pieces. The jig aligns the base plate of the biscuit joiner with the inside edge of the corner piece. If the corner piece is aligned correctly in the jig, the joint between the baseboard and the corner piece will be flush. It's important to cut the biscuit slots perpendicularly to the end of each corner piece; if the slots are skewed, gaps at the joints are inevitable. After marking the corner piece for slot locations, the cuts can be made on both edges.

With that done, we dry-fit the corner pieces to scrap pieces of baseboard and then use a router and a roundover bit on the top edges of the corner pieces so that they will match the top-edge profile of the baseboard. Once this profile has been cut, the corner pieces are ready for installation.

Making Corner Pieces on a Shaper

If there's not a mill shop in your area that will make the corner pieces, you might try to make them yourself. The required tools are a table saw, a shaper, and, of course, the necessary shaper knives. I wouldn't recommend a router because the bits required for the cuts would be very large.

A mill shop in my area makes all the corner stock I need. But if I were going to make my own corner stock, I'd use a three-step process on the table saw and shaper. First, I'd rip the stock to the appropriate dimensions. These cuts would establish the two faces of the corner piece that will be joined with the straight runs of baseboard.

Next, I would cut the inside curve on a shaper. The radius depends on variables like drywall or plaster thickness and the thickness of the base itself. Then I would go back to the table saw to rip the opposite (outside) corner to remove most of the waste material.

Finally, I'd use the shaper to finish up. Again, the radius required may vary. I think two passes would do it, each pass cutting half the outside radius. A little sanding will finish the job up nicely. If you try this, use material that's long enough to be machined safely.

USING A TABLE SAW AND A SHAPER

After the stock has been dimensioned and squared, use a shaper to remove material on the inside corner (1). Next, use the table saw to remove waste on the outside corner (2). The last cuts on the shaper (3) produce the outside radius.

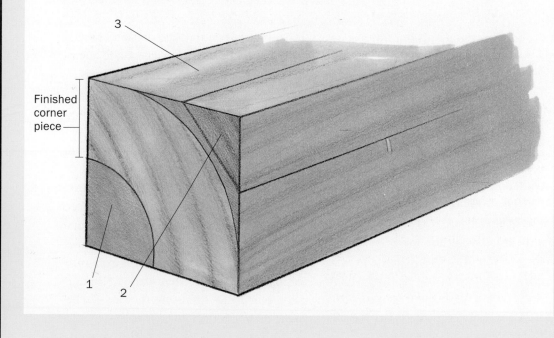

Finished corner piece

To Install Baseboard, Dry-Fit Corners First

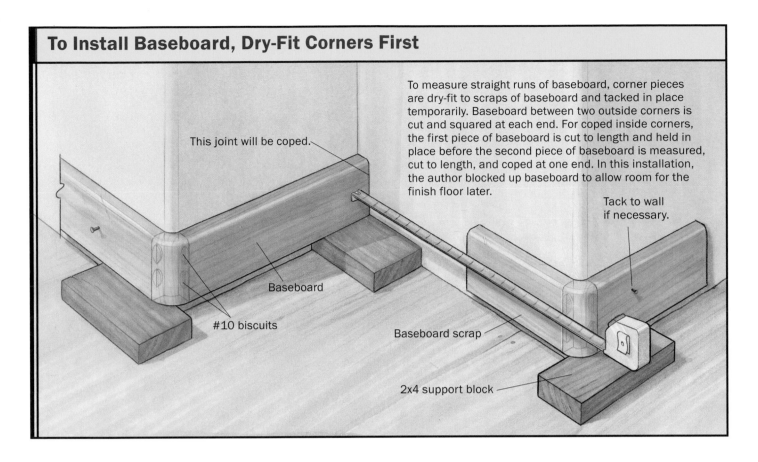

This joint will be coped.

To measure straight runs of baseboard, corner pieces are dry-fit to scraps of baseboard and tacked in place temporarily. Baseboard between two outside corners is cut and squared at each end. For coped inside corners, the first piece of baseboard is cut to length and held in place before the second piece of baseboard is measured, cut to length, and coped at one end. In this installation, the author blocked up baseboard to allow room for the finish floor later.

Baseboard

#10 biscuits

Tack to wall if necessary.

Baseboard scrap

2x4 support block

Measuring and Installing the Baseboard

The second step in the process is to set corner pieces in place temporarily and measure the straight runs of baseboard. In the Santa Rosa house, a tile floor was to be installed after the baseboard was installed, so we raised the baseboard 1½ in. off the subfloor and took our measurements for the base at this height (drawing, above). To hold corner pieces in place while we measured straight runs of baseboard, we dry-fit scraps of base to corner pieces with biscuits. The assemblies could be tacked to the wall to hold corner pieces in the correct position while we took measurements.

With the straight runs of baseboard all cut, we could install all of the pieces. At each outside corner, we dry-fit both pieces of base to the corner piece to check the

joints. If the fit looked good, we glued the slots, inserted the biscuits and then stuck the pieces together. If the dry fit had been perfect, we would have nailed the pieces in place right after the glue was applied.

If extra pressure were needed because one of the joints was slightly off, we let the joint set up off to one side before installing the pieces. When the glue was set, we nailed the assembly in place. It was sometimes possible, especially with shorter pieces, to assemble a three-piece corner or a five-piece U-shaped section, let the glue dry, and then install it in one piece. The process may sound tedious, but it's not. Once a rhythm is established, the work flows smoothly.

Eric Blomberg is a carpenter with Jim Murphy & Associates of Santa Rosa, California.

Running Baseboard Efficiently

■ BY GREG SMITH

Let's face it. There is little or no glory in the installation of baseboard. If you want, for instance, to talk about hanging doors, you can probably find plenty of guys who are happy to grant you their expert opinions on the best tools and the most elaborate techniques. But when it comes to installing baseboard, we're back to grabbing a scrap of lumber or an unspent napkin from lunch to record measurements.

The job may go something like this: enter room, plop saw on floor, measure, cut, nail; measure, cut, nail; measure, cut, nail. And you wonder if you'll ever get to the last piece, because it seems like there is always another little piece in some nook or cranny or some space that was missed. It is a job that brings screaming protest from the knees and a hacking voice of discontent from the lungs of the person who fires the nailer that kicks up the dust from the floor adjacent to the workpiece. That may be why, when a team of carpenters is finishing a house, running baseboard is often relegated to the least-experienced person of the group or the low man on the totem pole.

The best way to deal with an unpleasant, though necessary, task is to get it done as quickly as possible. I have seen many carpenters approach the installation of baseboard in many different ways, but I had never seen a system that works very efficiently. That's why I developed a methodical approach that makes baseboard installation fast and efficient.

1. Plan Your Strategy

The time to run baseboard is before the painter has hidden the location of studs (assuming that you are dealing with drywall) and after the door casing, the built-ins and cabinets, and the hardwood or tile floors are installed. In the areas that will be carpeted, hold the baseboard off the floor with a piece of hardwood flooring or other scrap of ¾-in. material—you won't want your beautiful work hidden by the carpet. Though I like to leave a wake of completed baseboard behind me when I am working, that's not always possible. If the bathroom floors have yet to be tiled, for example, I cut my baseboard for the room and set it aside.

2. Set Up the Saw

I usually set up my 10-in. power miter saw across extra-tall sawhorses (so that I don't have to bend far to see a close-up of the cut) in the biggest room on the floor level on which I'll be working. I'm not as concerned about how close I am to the area to be worked on as I am about having enough room to extend long lengths of material on either side of the saw. With my system, I don't spend a lot of time walking back and forth between my saw and the work area. If there are no large rooms, or if for some reason I cannot use one, I set up instead where I can extend stock out through a door or a window. When all of the baseboard material is spread out near the saw, I'm ready to start work.

By the way, no matter how clean the subfloor is, when you're shooting in baseboard with a nailer, the dust is going to be flying, and your mouth and nose, being close to the ground, are going to scoop up a lot of it. You might want to use a dust mask for this part of the job. In situations where I can free up my left hand, I put it under the place where air is released from the nailer. This keeps the dust from being kicked into the air.

As for what joints to cut, it's up to you whether you miter or cope. The system I use to organize the process won't change the way you work. For purposes of explanation, however, I'll use the example of mitered baseboard.

3. Install the Long Boards

Tackle the longest walls first—the ones that are longer than the stock you are using. These walls will require the baseboard to be spliced somewhere along its length (drawing, p. 54). Start by cutting lengths of stock with a 45° inside miter on the left-hand side and an inside 22½° miter on the right-hand side (assuming you're working from left to right). The splice will occur at the second

To run baseboard efficiently, Greg Smith measures several rooms at a time. He prepares a cutlist as he goes and marks lengths and cut angles on a preprinted form.

cut. Later on, you'll be able to take one measurement from the 22½° end to the corner of the wall to complete the wall. Go ahead and install these lengths of baseboard.

4. Take Closing Measurements

Now take all remaining measurements for three or four rooms at a time. Starting at the door, measure each and every length in the room. Work your way around the room in a clockwise or counterclockwise direction, whichever you prefer. The direction you

Cut Angles

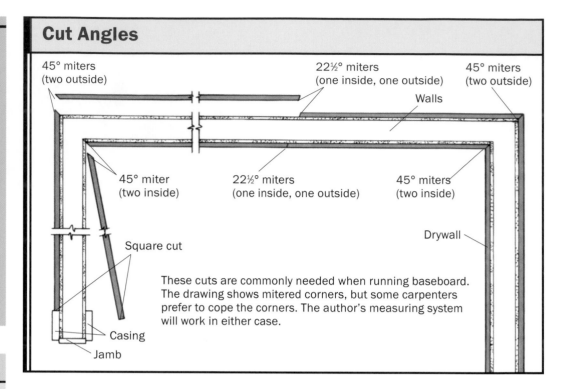

45° miters
(two outside)

22½° miters
(one inside, one outside)

45° miters
(two outside)

Walls

45° miter
(two inside)

22½° miters
(one inside, one outside)

45° miters
(two inside)

Drywall

Square cut

These cuts are commonly needed when running baseboard. The drawing shows mitered corners, but some carpenters prefer to cope the corners. The author's measuring system will work in either case.

Casing

Jamb

choose is not as important as being consistent. Use a mechanical pencil (the kind you can get in any grocery or drug store) both for recording your measurements and for marking cutlines later on. The point will always be sharp and consistent, and you won't whittle away time carving up a carpenter's pencil with a razor knife.

Each measurement is recorded on a very simple form that I've designed and carry on a clipboard. The column of blanks is sequentially numbered, and the measurements are recorded in order. The sequential numbers are important. They will later be recorded on the boards and used to guide you toward correct board placement. You can increase your ability to keep track of what you're doing by drawing a line between each set of measurements when you change rooms. Write the name of the room in the vertical space to the left of the column. In the example, boards #1 to #4 go in the master bedroom, boards #5 and #6 go in the closet.

The solid line to the right of the blanks represents the baseboard as you are looking at it on the wall. On the left and right ends of this line, you will write a symbol to repre-

sent the kind of cut required on each end. Use whatever symbols make sense to you—but be consistent. In the small example on the facing page, the straight, vertical line on baseboard #1 means the cut is to be a straight, square cut. This end of the board will butt against the casing around a door.

The "2" on the right-hand side of the notation represents an inside 22½° cut. Baseboard #2 starts with a 22½° cut; the slash at the end of the baseboard line indicates a 45° inside cut. Baseboard #4 shows a square cut and a 45° outside cut (represented by the O at the end of the line). When working with a stain-grade wood, such as oak, most carpenters like to cope one end to get a tighter fit. I use a C to indicate the end to be coped. You won't find many different cuts in baseboarding, so it won't be hard to memorize the symbols you'll need. If you work with others, you might want to define your symbols on the form so that everyone will be singing from the same song sheet. As for measuring, I simply hook the tape where it's most convenient.

Baseboard Cutlist

The sample below shows how to use the cutlist (right). Marks at the end of each solid line indicate the cuts to be made on each baseboard.

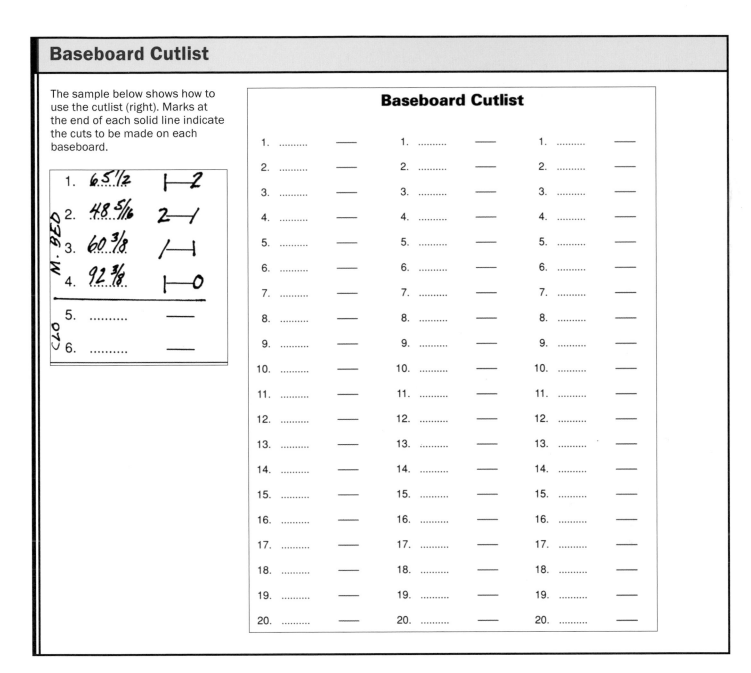

5. Cut Each Closing Board

After you have compiled measurements from several rooms, take your clipboard to the saw and start cutting baseboard. Each time you cut a board, mark the backside near an end or in the middle (be consistent so that you will know where to look for it later), using the sequential number on your list. Then cross that length off your list. There is no need to write exact lengths on the board, as many carpenters do. For example, if you cut a 22½° inside cut on the left-hand side and 107⅜ in. to the right, make a 45° inside cut; turn the board over, write the number 3, and cross it off your list. Then set the board aside and begin work on board #4.

You need not cut in any particular order. One additional advantage of this method is that you can make very efficient use of your material by taking a little extra time here to avoid waste. Start with your longest pieces, then see what you can get out of the offcuts. You'll have a long list from which to choose.

With a pile of stock nearby and his list of baseboard dimensions at the ready, Smith parks himself at the saw and cuts down the list. Each piece gets a number that corresponds to the list; later on, he'll be able to mate each length with the correct wall.

6. Distribute the Boards

Now that you have cut and numbered all of the pieces on your list, you can distribute the pieces. Because you have numbered them in the order in which you measured the walls, it is easy to pick up any random piece and quickly find the place where it belongs. Let's say that the first board you pick up is marked 18, and you see that it is the third board cut in the dining room. You can go to the dining room immediately and set the board alongside the third wall from where you started measuring in that room.

7. Nail 'Em Up

When all the pieces are lying in front of their respective locations, it is time to crawl around the room and nail them up. You may occasionally find that a board or two will need to be recut or trimmed because of walls being out of square or because you

measured wrong. This would happen regardless of your approach to the job and will not affect your other cuts. You can either recut as you come to them or write the adjustment needed on the back of the board; e. g., "⅛" indicates that you need to cut ⅛ in. off of the length.

This system also works well if you like working with a partner. One person measures and installs, and the other does all of the cutting. Work this way with a partner only if you like and respect each other and if each of you has a sense of humor. Inevitably there will be discussions about who can't measure right and who can't cut right when boards occasionally come up short or long.

Nothing Fancy, but It Works

It's a simple system, really, but it does work. You don't have to use a preprinted form, of course, though it does eliminate the need to write out the sequential numbers and draw lines. Besides, it's a heck of a lot easier to write on and read than some of the things that you see carpenters writing on at construction sites.

To save you a bit of setup time, you can enlarge and photocopy the sample form on p. 55 for your own use. Photocopy enough to keep yourself supplied for a few months and keep a few extras in your truck. The columns will allow you to work through the "measure, cut, nail" process 60 times per page. You can also use the form to record measurements for other types of molding and for closet poles. Baseboarding may still be the lowliest job going, but with a bit of organization, you won't be down there quite as long.

Greg Smith is a general contractor in West Los Angeles, California, who specializes in custom home building and remodeling.

Designing and Installing Baseboards

■ BY JOSEPH BEALS III

Baseboards are so called because they derive from the base of a pedestal, the lowest part of the stand upon which classic columns are placed. Almost all the interior trim in a home is modeled on classical antecedents: Many great period houses reproduce classical models in every magnificent detail, while less elaborate trim echoes the form with a range of derivative elements. Column capitals and the entablature above become frieze and cornice of ceiling molding, and columns themselves become door and window casings. Pedestals are represented by a paneled wainscot under a chair rail: When the paneling is eliminated, the baseboard remains.

The clamshell (or ranch) base in a tract house shows how the economies of construction have reduced baseboards to a thin shadow of their former grandeur. But with simple shop detailing and stock molding, handsome baseboards can be built in a variety of designs and styles.

First the Floor, Then the Baseboard

Wooden floors, particularly strip and plank floors, require room to move with seasonal changes of humidity. A small gap between the flooring and the wall allows this movement, and baseboard covers the gap. A tight joint between the floor and the baseboard requires that the floor be installed first, and the baseboard scribed to it.

In a room to have a wall-to-wall carpet, the baseboard should be installed first, laid to the subfloor without scribing. It isn't necessary to tuck carpet under the baseboard: Carpet installers will fit the carpet tight against it, hiding the bottom edge entirely. Sometimes, narrow-stock baseboards, such as clamshell, are installed ½ in. or so above the subfloor, but that's only to keep the carpet and pad from obscuring too much of the shallow profile.

Scribing ensures a good fit between baseboard and floor. With the baseboard shimmed up parallel to the floor and about 1 in. above it, the author uses his scribe to trace the contour of the floor along the bottom of the baseboard.

Layout Lines Ensure that Scribed Baseboards Line Up around the Room

In its simplest form, a baseboard is really a foundation for the more detailed profiles that can be developed from it. But a simple baseboard is an ideal model for reviewing installation. This design calls for a 1x8 baseboard with a stock cap.

I start by marking a pair of reference lines along the walls, parallel to the average floor plane. The lower line represents the top edge of the installed 1x8. The lower line must therefore be no more than 7¼ in. (the width of stock 1x8) above the floor at any point around the room. In practice, setting the lower line at 7 in. ensures a tight fit in case you miss a dip in floor.

I use a long straightedge to establish the line location, but I do not make the mistake of working to a level reference. Baseboard is effectively floor trim, and the floor plane is the critical reference, level or not. Making the baseboards level in a room that is not level merely draws attention to the deficiency.

The upper line is a scribe line, parallel to the baseline and 1 in. or so above it. An inch is usually sufficient, but for a really erratic floor, you must ensure that the space between the two lines is greater than the maximum variation in floor height.

Once I've located the base and scribe lines, I snap them along all walls with a chalkline. You may think this is far too complicated for baseboard work. But the layout is not difficult, nor time-consuming. It may be possible to avoid it if simple base is being fitted in new construction, but the antique houses where I do most of my work never allow that degree of convenience.

Why Not Use Shoe Molding Instead of Scribing?

A shoe molding is a low-profile trim piece—like quarter-round, but a little taller—nailed along the bottom of the baseboard. Shoe moldings have been specified for their looks alone, but their origins are undoubtedly functional, where they cover some defect in the joint between the baseboard and the floor. A shoe molding is generally not a good alternative to scribing, because it will follow only gentle curves. The molding itself is much too small to scribe, so it cannot be used on a highly irregular floor with any success.

Good design requires that a shoe abut the sides of the plinth blocks (sidebar, p. 62) along with the rest of the baseboard. In most cases, though, a plinth block thick enough to accommodate the addition of a shoe will be aesthetically impractical. It's common practice to dead-end the shoe with a simple bevel or a mitered return. This condition alone is enough to make the shoe look like a badly planned afterthought.

Fit One End, Scribe the Bottom

I like to start with a short, inconspicuous piece of baseboard, if there is one, to get focused on the demands of the job. Before I can scribe the bottom, I need to fit one end. Otherwise, shifting the baseboard horizontally to fit an end will ruin the accuracy of the scribe. The shift may be only a small fraction, but on an irregular floor that will spoil the scribed fit completely.

If my first piece is trapped between two walls, the end cuts are not critical because the abutting baseboards will cover them. But if I must fit one end to a plinth or to a piece of installed base, the end cut needs to be perfect. I hold the base as low as possible, keeping it parallel to the scribe line. I can't

use the baseline, because until I've removed the scribe waste from the bottom of the baseboard, the baseline will be hidden behind it. I scribe the critical end and remove the waste with a jigsaw. Then I trim to the line with a sharp block plane at a slight angle, perhaps 5°, to relieve the back of the joint.

Now I hold this run of base tight against the end I've just fitted, aligning the top edge on the scribe line. If the untrimmed end runs past an outside corner or over a plinth block, I leave this several inches long. I'll scribe right out to the end and trim it afterward.

I shim each end of the base so that its top edge lies exactly on the scribe line. Then I set my scribe, or compass, equal to the distance between the two layout lines.

I run the scribe along the floor (photo, facing page), being careful to keep the two legs in the same vertical plane. I set the baseboard across sawhorses and remove the waste with a jigsaw, tilting the shoe to relieve the back of the cut (top left photo, p. 60). Because this is hidden, the relief can be strong, perhaps 10° to 15°. I cut close to the line, but never into it. After sawing, I trim tight to the line with a sharp block plane (top right photo, p. 60). I do not angle the plane; by keeping the plane perpendicular, it trims only the sharp edge of the sawcut and makes the fine-tuning fast and accurate.

With one end already fitted and the bottom now tight to the floor, a long end that is to abut a plinth can now be trimmed. I mark the top of the cut off the plinth. A line on the floor, extended off the bottom of the plinth with a square, shows the bottom of the cut. I saw the end as described previously, saving the line. I attempt a trial fit before I plane the cut because sometimes the baseboard wants to spring home directly off the saw. Failing that happy circumstance, a glance at the fit before final trimming is good protection against trimming short (bottom photo, p. 60).

A shoe molding is generally not a good alternative to scribing, because it will follow only gentle curves.

TIP

A glance at the fit of the baseboard before final trimming is good protection against trimming short.

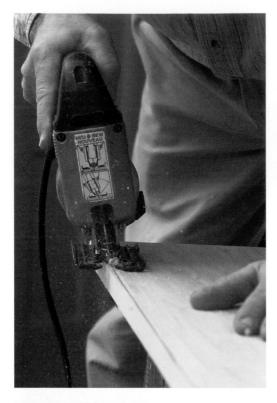

After tilting the jigsaw's shoe to create a back cut, the author cuts close to the scribed line but never into it.

A block plane fine-tunes the scribe. Having back-cut the baseboard with the jigsaw, planing (with the plane held square to the face) is easier because there's less stock to remove.

Glued-up plinths are more efficient. To save time and material, the author glues up plinth blocks two at a time, back to back. Then he rips them apart on the table saw after the glue cures.

Mitering an Outside Corner

An outside corner won't necessarily be square, nor will it necessarily be perpendicular to the floor. A square-cut, 45° miter almost always seems the right thing to do, and it almost never fits. Here's how I mark and cut outside corners for an accurate fit every time.

I fit either piece of base tight to the floor and to its far end and let it run a few inches past the corner. Then I hold a piece of base stock against the wall on the other side of the corner so that its bottom edge rests across the top of the first piece (photo, facing page). I draw a sharp pencil line along each side of the bottom edge, making a pair of marks across the top edge of the first piece. Then I draw one more sharp pencil line on the floor along the outside bottom edge of the first piece, just where it runs past the corner. If you don't care to draw on the floor, stick a piece of tape at the corner beforehand.

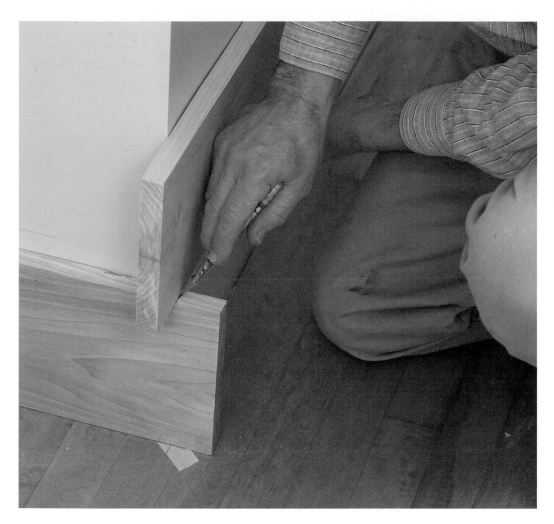

Mark baseboard in place when outside corners aren't square. Pencil lines on a piece of tape show where the baseboards intersect at the floor. A line connecting this point with the point of intersection at the top of the baseboard (being marked here) establishes the outside edge of the miter. Similar lines on the back of the baseboard determine the bevel angle.

Next I fit the second piece and mark it in the same way. When I mark a line along its outside bottom edge on the floor, this line will intersect the one drawn previously. I mark this point on the bottom edge of the second piece of base. Then I set the second piece aside, replace the first piece, and mark the same intersection on its bottom edge. The miter cutlines can now be drawn on both pieces by connecting all of the lines.

Unless the miter happens to be a 45° angle square to the floor, it isn't worthwhile setting a power saw to cut the line accurately. It takes too long to set up a compound-angle cut, and a mistake that spoils the piece can happen easily. Instead, I cut close to the line with a power saw or use a sharp handsaw: I leave ⅟₁₆ in. or so proud of the line and trim with a sharp plane. I trim the top edge accu-

rately to the diagonal line, but relieve the back of the miter below the top edge because this part of the joint will be hidden.

I test the fit and fine-tune the miter as necessary. Then I apply a bead of yellow glue to the miter, and nail both sides in turn, keeping the miter tight by alternating the nailing as necessary. Nail to the framing studs in the corner, and finally, nail through the miter with 4d or 6d finish nails to lock it up.

Nailing It All Home

Each piece of base should be nailed in turn because the fit of adjoining pieces depends on it. Nailing tall base can be a problem, especially in old houses where studding is often irregular. The bottom of the 1x8 base

Plinth Blocks and Baseboard Design

The first element in baseboard design isn't the baseboard, it is the door casing. Baseboards should sit at least ⅛ in. behind the face of the casing to create a crisp reveal because a true flush joint is almost impossible to make. Often, the best way to create this reveal is with plinth blocks.

A plinth block is just that: a block slightly wider and thicker than the casing and slightly taller than the baseboard. Once you are accustomed to plinth blocks, casings will look awkward without them, rather like legs with the feet missing.

Plinth blocks can be made of solid stock, but it is easier to use ¾-in. stock packed out with a pair of strips glued on the back. This avoids having to find thick stock, and the packing provides a relief that straddles any irregularities where the wall meets the door jamb.

Blocks can be made up in pairs with ¾-in. square strips interposed between them. The pairs are ripped apart after the glue cures (bottom photo, p. 60).

Once you've settled on the door casing, you can turn your attention to the baseboard itself.

The photos below and those on pp. 64 and 65 illustrate how you can easily alter baseboard design with stock moldings and simple shop millwork.

BASIC 1X8 WITH A COVE CAP

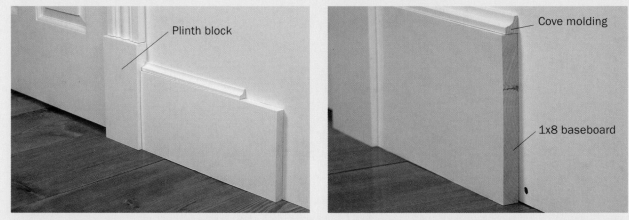

Although 1x4 or 1x6 may work fine, the height of 1x8 baseboard distinguishes it from the narrow baseboards common today. Here, the 1x8 is capped with a stock cove molding.

ADDING ½-IN. BEAD SHIFTS THE DESIGN TO SEMIFORMAL

The author makes this bead in his shop by resawing 2-in.-wide stock to ½ in. thick. He routs a bead on both edges and then rips two ⅞-in. wide beads from each piece.

can always be nailed into the sole plate. I use an ultrasonic stud finder when there are no obvious signs of the studding, but this doesn't help when there is no stud.

Wide door casings often extend past the rough framing. To fasten the top corner of the baseboard, an 8d or larger finish nail can be sharply angled to catch the stud behind the casing. Where this doesn't work, the nail can be driven through the top edge of the base, toenailing it into the side of the plinth. A bead of construction adhesive behind the base is a good way to back up marginal nailing and can be invaluable at inside corners. There is no practical way to pull the top of the baseboard tight to the wall between studs. But a gap between the baseboard and the wall can be hidden by cap molding.

Putting a Cap on It

I prefer to install the baseboard cap molding counterclockwise around the room because this puts the coped cuts at the right-hand end of each piece, which is easier for me as a right-handed person.

Coped joints have three distinctive advantages over miters: First, a coped joint will accommodate small variations off square with little or no fine-fitting; second, coped joints tend to stay tight even as wood moves with seasonal changes of humidity; third, a coped joint can be made tight even if there is a small variation in the profiles of the mating moldings, which is an unfortunate but not uncommon condition. By comparison, a mitered joint is a fussy creature that can do none of these things, and it has a bad habit of opening up when the mating pieces are nailed tight.

TIP

If you're right-handed, install the baseboard cap molding counterclockwise around the room because this puts the coped cuts at the right-hand end of each piece.

Coped joints work better than miters. Rather than mitering inside corners, coping (as shown here) works well even if the corner isn't square, hides inconsistencies in molding profiles, and stays tight with seasonal movement.

A SIMPLE BASE CAP CREATES A MORE CLASSICAL LOOK

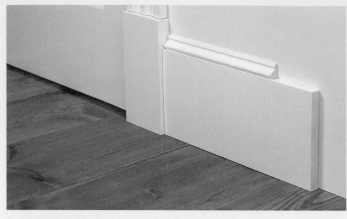

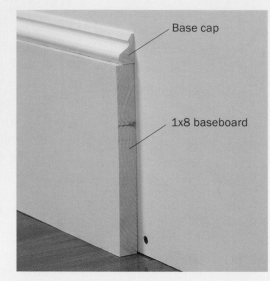

Base cap

1x8 baseboard

Substituting a simple base cap for the more angular cove molding shifts the baseboard's appearance toward a clean, more classical design.

A ½-IN. BEAD DRESSES UP THE BASE CAP

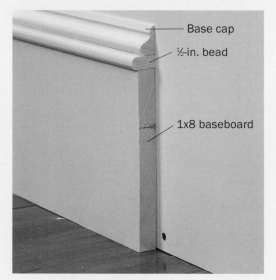

Base cap

½-in. bead

1x8 baseboard

Once again, interposing a bead between the baseboard and cap creates a more formal look.

AN ALTERNATIVE BASE CAP HAS SOFTER LINES

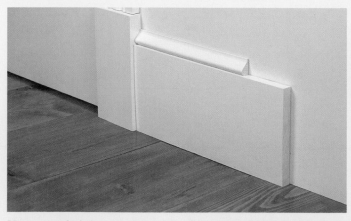

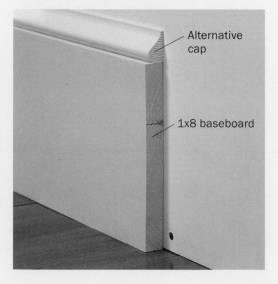

Alternative cap

1x8 baseboard

More rounded than the base cap above, this design is somehow softer in appearance. This profile is a stock molding but may be hard to find in this size.

A SHOPMADE CAP WORKS WITH BOTH COLONIAL AND CRAFTSMAN STYLES

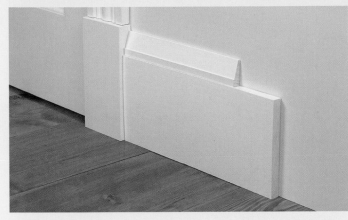

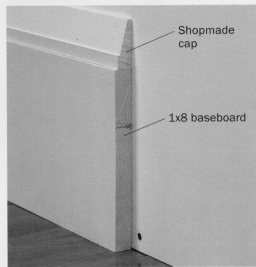

Shopmade cap

1x8 baseboard

This base cap was made on the table saw and was cleaned up with a handplane. The molding is 2 in. tall by ⅝ in. thick.

FURRING OUT THE BASEBOARD ALLOWS MORE FORMAL DESIGNS

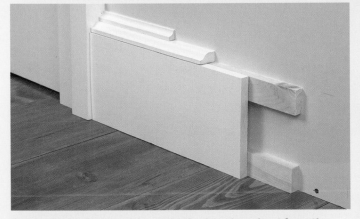

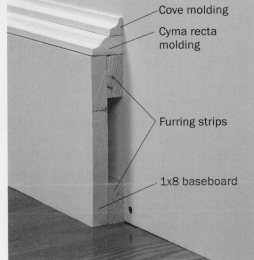

Cove molding

Cyma recta molding

Furring strips

1x8 baseboard

A pair of ¾-in. furring strips hold the baseboard out from the wall, creating space for a shopmade cyma recta molding between the baseboard and a cove cap.

EXTRA DEPTH MAKES ROOM FOR STILL MORE LAYERS

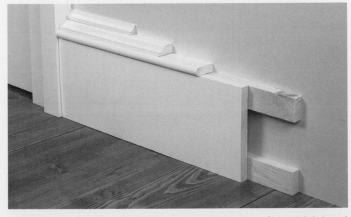

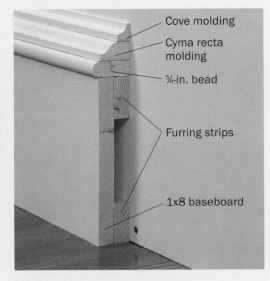

Cove molding

Cyma recta molding

⅝-in. bead

Furring strips

1x8 baseboard

With the deeper design, the author substitutes a ⅝-in. thick bead (instead of ½ in.) but cautions against the temptation of using full ¾-in. stock, which would look clumsy.

The first length of cap is cut square at both ends and fitted tight. The second length of cap is cut long with a miter on the right-hand end. I darken the front, profiled edge of the miter with the side of a pencil lead to highlight the line. Then I make the coping cut with a sharp, fine-toothed coping saw (photo, above), and hold the cap so that I can visualize the joint as I cut it. The coping cut is made square across the top, but immediately below the top the cut is angled to relieve the back of the joint. I saw accurately to the line if possible, and this may require approaching parts of the profile from both top and bottom. I use a half-round rasp and a coarse, rat-tail file to dress the cope tight to the line and to clean any parts too awkward to saw.

In a typical room, the last piece of cap must be coped at both ends. On a long length of cap, a spring fit will offer enough latitude to fine-tune each cope for a good fit, but short runs coped at both ends are notoriously difficult.

I nail the cap as I go (photo, facing page) so that each piece is fitted to a right-hand neighbor already nailed tight. Some carpenters argue about whether the cap should be nailed to the base or to the studs. My goal is a clean, tight fit, and I'll nail in whatever manner produces it. Where the cap spans a length of wall that is concave between two studs, toenailing into the back corner of the base will often pull the cap tight to the wall, or tight enough that caulking hides any remaining deficiency.

Priming Is Better than Plywood

Wide base stock that can be used to advantage with formal and classical caps may raise questions about wood movement, and you might wonder if hardwood veneered ply-

The base cap is nailed on last. Besides adding a decorative profile, the base cap conforms to irregularities and hides any gaps between the wall and the baseboard. The cap can be nailed either to the wall studs or to the baseboard.

wood is preferable to solid stock. I used birch-veneer plywood to make the tall baseboards in a small, paneled dining room. The results were excellent, but the plywood was not cost-effective. All runs in the dining room were less than 8 ft.; but longer runs would have required end joints, which are difficult to hide. All pieces must be gotten out of 8-ft. lengths of stock, which leads to a high waste factor. Trimming the scribe is tedious because the end-grain veneers fight the plane, and the thin face veneer is prone to breakage.

I've installed tall baseboards in solid stock without any problems, and I attribute this situation to the rigorous use of primer. I always back-prime baseboards with a top-quality, heavy-bodied oil primer, and I seal all ends.

For species, clear pine is too expensive. No. 2 pine, picked to avoid obvious defects, is inexpensive and finishes well. Poplar is a good choice. It's a relatively clear, fine-grained hardwood that works easily and takes paint well. Some lumberyards offer poplar as dimensional stock; for a better price, see a hardwood dealer.

All the designs shown here feature 1x8 baseboard stock, but 1x6 or even 1x4 may better suit your intentions or the room itself. If you want to experiment, prepare a trial run of baseboard and cap at least 2 ft. long that you can place in the room where it will be installed. Designs that look good on paper can look different on site, viewed from 5 ft. or 6 ft. away.

Joseph Beals III is a designer and builder in Marshfield, Massachusetts.

TIP

Back-prime baseboards with a top-quality, heavy-bodied oil primer, and seal all ends.

More Than One Way to Case a Window

■ BY JOSEPH BEALS III

Stock trim, such as 1x4 pine or clamshell casing, will always be useful in routine, contemporary construction. But after years of installing these lifeless, standard-issue casings, frustration drove me to cross the frontier. Using as a reference the casings I'd seen in so many period New England houses, I built a simple pediment head casing to surmount square-edge, square-cut side casings. I will not forget that first look back at the result: An ordinary window suddenly had character, grace, and purpose; and a client new to classical trim was particularly pleased with a window worth looking at, not just through.

Mitered Casings Get Moldings; Butted Casings Get Pediments

In this article, I'll discuss two approaches to formal casing design: mitered casings as a stylistic option and butted casings, which employ a pediment head, or architrave, as a classical architectural option. In general, mitered casings are developed by adding layers of moldings to the perimeter of a mitered flat casing. With butted casings, the pediment head sits on the square-cut tops of plain or molded side casings, and architectural detail is developed on the pediment itself.

In a mitered casing the simplest alternative is the use of two or more layers of molding. A bead cut on the inside edge of a casing and a backband applied around the outside perimeter of a flat casing will give a strong, three-dimensional appearance (top photo, p. 76). A thin molding interposed between the flat casing and the backband adds another element of detail and shadowline to the profile (bottom photo, p. 76).

A formal alternative to mitered casings is the use of a pediment, or architrave, as a head casing with side casings butting into it. The pediment represents an entablature, the lower portion of a classical roof, and its elements are derived from ancient Greek and Roman temples. This style of window trim is commonplace in period architecture, especially in Federal and Greek Revival houses of the 19th century.

Two Approaches to Formal Casing

TIP

When using a two- or three-piece mitered casing, figure out the combined thicknesses of the molding before you cut the stool.

With mitered casing the same trim runs up one side, across the top, and down the other. Butted casing entails a classical pediment sitting on top of square-cut side casings.

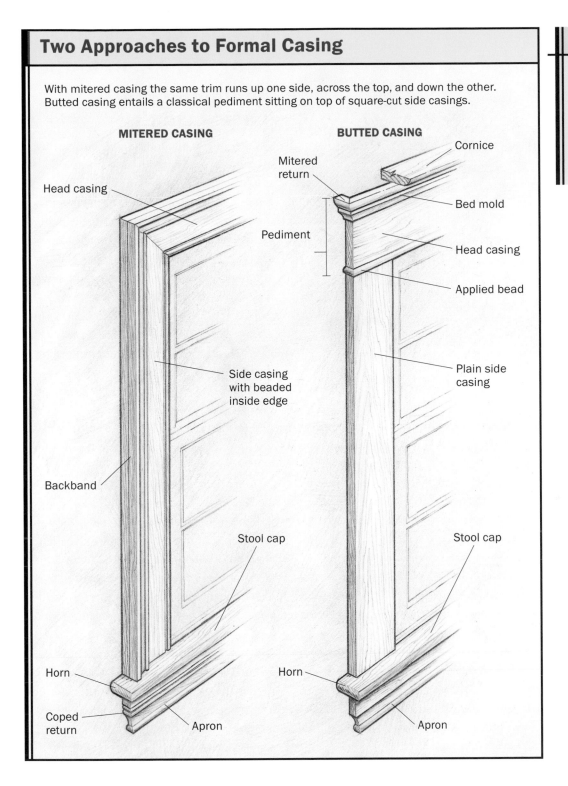

MITERED CASING

Head casing

Side casing with beaded inside edge

Backband

Stool cap

Horn

Coped return

Apron

BUTTED CASING

Mitered return

Pediment

Cornice

Bed mold

Head casing

Applied bead

Plain side casing

Stool cap

Horn

Apron

All Window Casings Start with a Stool Cap

Any style of window trim must begin with the stool cap, the piece that finishes the inside edge of the windowsill. You can buy stock stool cap at a lumberyard, but you'll spend enough time fitting it to the window that you might as well make your own.

I like the stool to stand proud of the casings about an inch. If you're thinking about a two-piece or three-piece mitered casing, you'll have to figure out the combined thicknesses of the moldings before you cut the stool.

Cut the Side Casings Long, Fit Them to the Stool, and Mark Their Length

For any casing design, the two side casings are cut long and fitted first to the stool. I cut the casing bottoms square, then hold a casing in position against the window jamb. I judge the reveal by eye (the reveal is the exposed portion of the jamb between the casing and the jamb's inside edge), but if you prefer reference marks, make them with a gauge block of some kind.

In theory, if the bottoms of the side casings are cut square and if the stool is perpendicular to the sides of the window, the casings should fit tight to the stool, but theory doesn't always work. I close any gaps by dressing the casing bottom with a sharp block plane. After both side casings have been fitted to the stool, I tack them in position with a few 4d finish nails.

To mark for the top cuts, I use a gauge block (top left photo, facing page). For mitered casings, the mark I make indicates the bottom of the miter. I draw a diagonal reference mark on the casing to remind me of the angle of the cut. This procedure may seem foolish, but it is remarkably easy to forget how things go together in the short walk to the miter saw.

When I mark the two sides of a butted casing, I make tick marks on the inside edges with the gauge block and then connect the two marks with a straightedge (bottom photo, facing page) and a sharp pencil. In theory, the top cuts should be square to the casings, but this theory may not be true in practice, especially in the case of an old window that you're retrimming. Before nailing the side casings, I cut biscuit slots in the bottom ends and mating slots on the top of the two stool horns (top right photo, facing page).

I cut the stool stock long, and I machine a bullnose, or half-round, on the outside edge. I scribe and cut the two horns (horns are the ends of the stool that overlap the drywall on each side), check the fit, then cut the ends to allow about a 1-in. overhang past the outside edges of the side casings. The bullnose returns can be shaped by machine, but I prefer to remove the bulk of the waste with a block plane and finish the job with a sharp file or a piece of cloth-backed sandpaper.

Given the wide variety of window conditions, fastening the stool can be a problem. On the Andersen℠ window used in these photos, the stool is toenailed from the top into the sill. The horns can be face-nailed into the framing studs if necessary. Glue or caulk between the stool and the sill can help hold everything together.

To keep the stool square to the window, I screw 2x4 blocks to the rough framing right below the stool (photo, above). Nothing is worse than getting a window trimmed and realizing that during the course of your work, you've pushed the stool out of square with the wall. The blocks hold the stool square until I add the apron at the end of the job.

Mark the side casings with a gauge block. The author uses a scrap of wood, marked with the amount of offset between the window jamb and trim, to mark the length of the side casings.

Common practice is to nail up through the stool into side casings. But biscuits hold better than end-grain nails and won't come out through the exposed surface of the side casing.

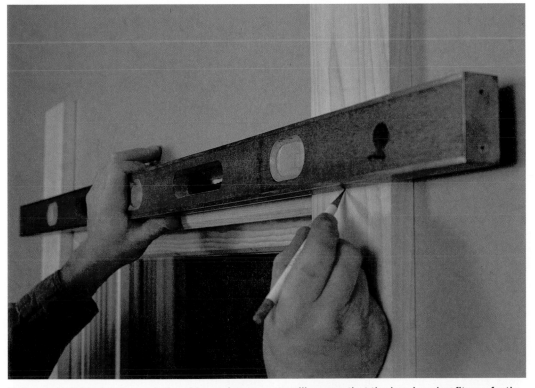

Using a straightedge to mark the side-casing top cuts will ensure that the head casing fits perfectly. Simply squaring cuts from the marks can result in sloppy joints.

Nail the Side Casings Home before You Fit the Heads

Some people prefer to fit the head casings while the two side casings are only tacked in position on the theory that it's easier to make the joints perfect if both pieces are adjustable. Then, after everything is fitted, you nail the pieces home in one marathon effort. I think this process invites problems because the side casings can shift as they are pulled against the wall when the nails are driven home. Even setting a nail after it has been hammered in can cause the casing to shift, and the joint you thought was perfect opens up again. Adjustments to bad joints are awkward or impossible.

After I fit a side casing to the stool, mark it, cut it to length, and cut the biscuit slot, I nail it tight and set all the nails. The side casings now are finished; they have been locked in place so that they can't move.

Other people prefer to install mitered casings by working around the window: up one side, across the top, and down the other side. I prefer my method because the two

sides are done quickly and easily, and any adjustments are made in the head casing. Only if the window is badly racked would this be impractical, in which event it might be better to fix the window condition first.

Fitting a Mitered Head Casing

I like to cut a bead on the inside edge of all of my mitered window casings. A shaper, router, or cutter head-equipped table saw cuts beads equally well. The molded bead mimics the applied sash stop found on a lot of old double-hung windows.

For a mitered head casing, I make the 45° cut on one end and then mark the other end. The casing is held upside down with its mitered end perfectly registered on the side casing, and the other end is marked with a utility knife (photo, below).

I cut the second miter, saving the knife line. I drop the head casing in place, taking care to keep it parallel with the window head jamb. Any deficiencies in fit are apparent, and these flaws are dressed out with a sharp block plane.

I use biscuits to reinforce the mitered joints. The biscuit slots in the side-casing miters can easily be made in place if they weren't done earlier, and the head-casing miters can be slotted on any flat surface. If all is well, I glue the slots, slide the biscuits in, and tap the head casing into place.

More Layers Add Depth and Detail to Mitered Casings

As the photographs of finished casing styles show, mitered casings can be built up in layers to give a handsome appearance. The tight, solid base offered by the flat casings makes the application of additional components easy.

For mitered casings, nail home the two sides and then fit the head. The author makes a 45° cut on one end of the head casing, holds it upside down on the side casings, and marks for the other cut with a utility knife.

It's possible that the outside edges of all of the built-up moldings of a mitered casing won't be perfectly flush after they're nailed on. A couple of passes with a bench plane makes a flat surface. The last few inches at the bottom are cleaned up with a sharp chisel.

There are plenty of backbands available at lumberyards, or you can make your own with a table saw and router. I install the backband in a sequence similar to the flat casings: fitting and nailing both sides and then fitting and nailing the head piece.

A three-part casing with an intermediate molding between the backband and flat casing adds another level of detail to the casing. I make the intermediate molding from ½-in. or thinner stock. Making this piece from thicker stock can result in a clunky, heavy casing. A thinner intermediate molding has a more delicate look.

On these multilayered casings, the outside edges may need dressing to present a single, flat surface. I do this with a sharp bench plane (photo, above), working as close as I can to the stool horns, and I clean up the last few inches with a sharp chisel.

Pediments Dress Up Butted Casings

One of the beauties of adding a pediment to a butted casing—aside from the aesthetic ones—is that all of the pediments can be made on a bench and added as a complete unit to the tops of the two side casings.

A basic pediment is built from 5/4 stock, with an applied bead across the bottom and a bed molding across the top (bottom left photo, p. 77). The space between the top of the bead and the bottom of the bed molding should be at least equal to the width of the side casings. You can make the space wider for a bolder appearance, but beware of overdoing it. Make a trial pediment if there is doubt about the aesthetic effect.

> **TIP**
>
> *The space between the top of the bead and the bottom of the bed molding should be at least equal to the width of the side casings.*

Six Variations on a Theme

We asked designer/builder Joseph Beals III to show us how to create various styles of window trim without resorting to custom moldings made with a shaper. On a test wall in his shop, Beals mocked up six casing designs, using mostly dimension lumber and stock moldings from the local lumberyard. He cheated occasionally and used his shaper, but where he did, a router would work just as well.

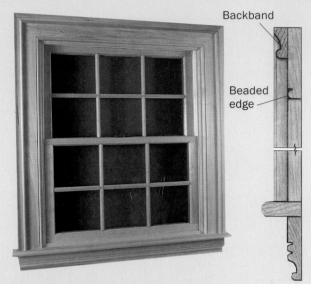

Backband

Beaded edge

MITERED CASING WITH BACKBAND AND A BEADED INSIDE EDGE
Starting with 4½-in.-wide flat casings, the author mills a simple bead on the inside edge and runs a backband molding around the perimeter.

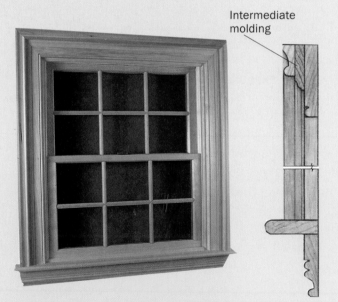

Intermediate molding

A THIRD LAYER ADDS DEPTH AND DETAIL
Interposed between the flat casing and the backband, a thinner (½ in. or less) intermediate molding adds another level of detail and shadowline.

SIMPLE BUTTED CASINGS

Plain 4½-in.-wide side casings are topped with a 4½-in.-wide head casing as a pediment. The only embellishment comes in the fact that the pediment is made from thicker stock (5/4) and overhangs the side casings in front and on the ends.

5/4 head casing

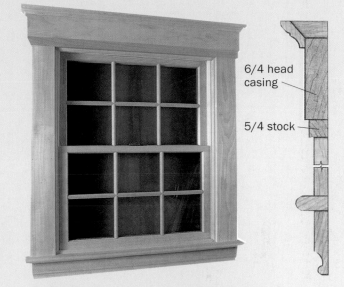

LAYERS OF SUCCESSIVELY THICKER STOCK LOOK ELEGANT

Here, the ¾-in. side casings support a pediment composed of 5/4 square stock and 6/4 head casing, again capped with a bed molding. Each element overhangs the face and end of the element below.

6/4 head casing

5/4 stock

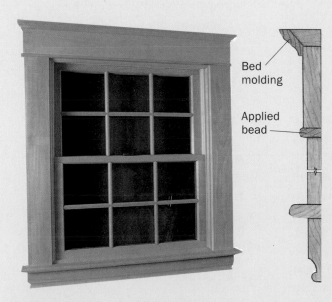

BED MOLDING ABOVE AND A BEAD BELOW DRESS UP THE PEDIMENT

A slightly wider head casing provides a nailing base for standard bed molding at the top of the pediment. Bullnosed on one edge, a thinner molding (⁷⁄₁₆ in. by 1⅛ in.) overhangs the bottom of the head casing.

Bed molding

Applied bead

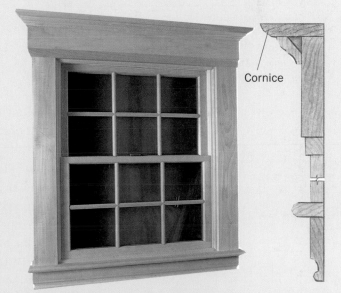

A CORNICE IS THE CROWNING TOUCH

Adding a simple cornice piece above the bed molding and head casing completes the basic elements for a formal entablature.

Cornice

Run mitered returns long then trim them with a handsaw. The author uses one hand to hold the handsaw while he makes the cut.

To align the pediment's components, the author nails the pieces together on a flat surface.

To provide a base for nailing on the bed molding, I make the height of the head stock at least ½ in. taller than the width of the side casings. The length of the head stock is equal to the distance between the outside edges of the side casings.

I draw a pencil line along the length of the head stock to indicate the bottom edge of the bed molding. With a square, I mark the bottom edge of each bed-molding return around the corner at each end of the head. This step is important because it's easy to cock a short return, and even a small misalignment will be brutally obvious once the pediment is installed.

The long piece of bed molding is mitered at each end. For convenience and safety, I cut the mitered returns long. I check the miters for a tight fit, then fasten the returns with glue and a few brads. To avoid nail holes, you can use masking tape to secure the returns while the glue dries. But this task is a slippery, three-handed job, and it won't make much difference in the final result. I use a handsaw to trim each return flush with the backside of the head (top photo).

To make the bead, I resaw a piece of stock $\frac{7}{16}$ in. thick by $1\frac{1}{8}$ in. wide, but these dimensions are not critical. I cut a bullnose along one edge with my shaper (also an easy router job). The bead overhangs the casings by about ⅜ in. at each end. I shape the returns with a block plane, then finish the radius with a sharp file or a strip of cloth-backed sandpaper. The bead is applied to the head stock with glue and 4d finish nails. You can do the application freehand, but working on a flat surface like a table saw makes it easy to keep the back of the bead flush with the back of the head stock (bottom photo).

Before installing the pediment, I cut biscuit slots on each side of the bottom, aligned with the centerlines of the two side casings. The slots in the side casings can be made before the casings are installed, but if you have enough ceiling height, it's just as easy to make them in place. Be aware of where you nail the bead onto the head stock so as not to put nails where they will interfere with cutting biscuit slots.

Pediments Can Range from Simple to Complex

As the photographs of casing styles show, the pediment can be varied to increase the level of architectural detail. The simplest pediment is a plain piece of 5/4 stock, cut to a length that overhangs the side casings at each end, as the bead does on the pediment described on p. 76). Adding a bed or cornice molding to the 5/4 stock begins to echo the lines of a true entablature.

The next evolution is a 6/4 head with a bed molding, surmounting a square-cut 5/4 fascia or lower molding (top right photo, p. 77). Each element overhangs the face and ends of the element below, in the same style as the classical Greek and Roman entablature.

Adding a cornice molding above the bed molding in any pediment completes the basic elements for the full entablature (bottom right photo, p. 77). For interior cornices, I use 4/4 or 5/4 stock, and I machine the curve on my shaper with a knife I ground for the purpose. I shape the returns by hand, as described above for stool and bead returns, because machine work on short, end-grain sections is usually awkward.

If you choose to include a cornice, you should increase the height of the head stock by the full height of the bed or cornice molding and install the cornice molding first. Remember to make the cornice long enough to incorporate the bed molding returns. Install the bed molding tight to the cornice, and take care to seat it properly. It's easy to cock the long molding in rotation, which will make fitting the mitered returns frustrating.

Aprons Complete the Casings

An apron is nailed to the wall below the stool and is the lowest component of a window casing. On a lot of windows, a piece of casing stock serves as the apron, with its ends cut square, angled, or returned, according to the whim of the carpenter or designer. There is no equivalent to the apron in classical architecture, which is why aprons in neoclassical window trim exhibit such a variety of profiles.

The apron has a specific aesthetic function, and there are several profiles that I use (top right photo). The apron visually returns the window to the plane of the wall below. About 3½ in. of height works well; if the

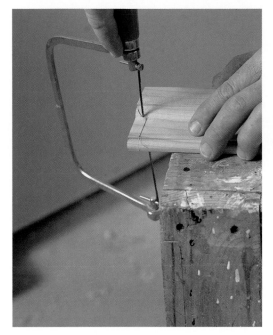

Rather than miter returns, the author draws an outline on the molding profile and cuts it with a coping saw.

Different aprons fit different casing styles. The author adds the apron after the rest of the window is cased, experimenting with the apron design to find one he thinks works best with a particular window-casing style.

apron stock is wider, it will look boxy and unbalanced, but narrower sections can be successful. I mill the profiles on the shaper and table saw, but a router will also handle this job easily. The simplest pattern echoes the backband that is used on the mitered casings. I make this piece out of solid stock, but a cove cut with an applied bead on the bottom is a simple alternative.

The apron returns are coped rather than mitered. I sketch the profile freehand, but you might prefer to use a paper or flexible plastic template. I remove the waste with a coping saw (top left photo), and I dress the return with a sharp file. This work goes quickly in contrast to mitered, glued returns, which are awkward and generally are not worth the effort, even on bright finished trim.

Joseph Beals III is a designer and builder in Marshfield, Massachusetts.

TIP

If you don't want to sketch the apron return profile freehand, use a paper or flexible plastic template.

Making Curved Casing

■ BY JONATHAN SHAFER

The curved casing for these windows was fabricated on site, using two pieces of straight stock cut into narrow strips and laminated around a form.

A few years ago I was asked to step in and complete the finish work on an 11,500-sq.-ft. Tudor home that had taken 18 months to get through drywall. Completing the trim took an additional 12 months and posed many challenges, such as hanging 8-ft.-high doors, building four stairways, and running thousands of feet of wide casings and base. The house also had many arch-top windows and doorways of various heights and widths. The curved window and door casings had to match the existing straight casing, so I decided to produce the curved casings on site with the help of a talented crew of finish carpenters.

My approach to this challenge was to strip-laminate the arched casings. By alternating strips from two pieces of straight, even-grained casing, we reproduced the casing profile. We ripped the strips from straight casing and then bent them around a form for each window and door. We also laminated extension jambs for each window, using the same bending forms.

Making the Patterns

Our first step was to make patterns of all the arched windows and doors. How we produced the patterns varied depending on the particular application—some methods were as simple as tracing on kraft paper (available in long rolls) against the window frame, while others were as involved as mathematically computing arcs and multiple radius points.

One method we used on some of the more complex windows required a thin, flexible ripping of even-grained wood long enough to follow the arch along a window frame. This strip was clamped or held by helpers against the inside of the frame. We maintained the arch shape by tacking crosspieces to the bowed strip. The more crosspieces we used, the better the shape was held after the clamps were removed. We then transferred the shape of the arch to kraft paper.

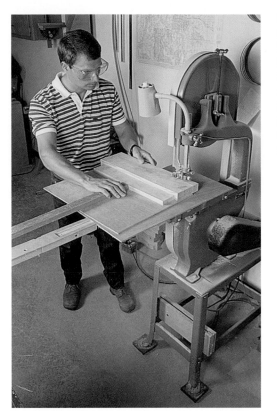

For round-top windows with small radii, Shafer cut sections of the bending forms on the simple jig shown in the drawing below. For bigger windows, he attached a length of 1x3 to the plywood carriage and extended it across the room to a center point on top of a workbench.

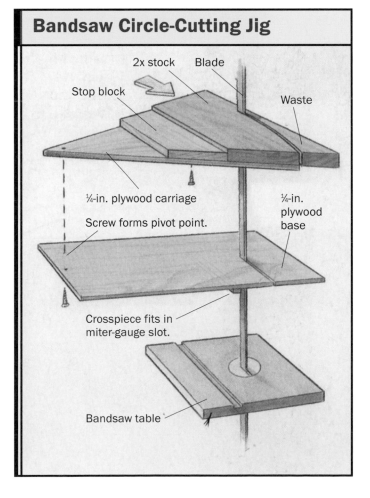

Bandsaw Circle-Cutting Jig

2x stock

Blade

Stop block

Waste

¼-in. plywood carriage

Screw forms pivot point.

¼-in. plywood base

Crosspiece fits in miter-gauge slot.

Bandsaw table

The basic principle here is that you're taking a piece of straight casing with the molded profile you want and ripping it into narrow strips that you can bend around a form and glue back together.

With another method, we tacked plywood against the window, using a piece wide enough to contain the unknown radius points. We then used a beam compass to find the radius points on the plywood by trial and error. Again, the arch was then transferred to kraft paper. I came on the project too late to have done it, but in the future I would make tracings of each window frame prior to installation.

Finally, we cut out each pattern and checked it against the corresponding window, making necessary adjustments. The patterns also had to be extended on both ends to allow extra casing length for trimming later. We labeled the patterns for window location and wood species.

Building the Bending Forms

When the patterns were ready, we built a bending form for each one. We constructed them for 2x stock cut into arcs on a bandsaw (for the design of a bending form for curved jambs, see the sidebar on pp. 82–83). With roundtop casings, the 2x arcs were made using a simple circle-cutting jig fixed to the bandsaw table. We extended the table with a piece of ¼-in. plywood and ran a screw through it to create a pivot point. The 2x stock was then pivoted around the pivot point on a ¼-in. plywood carriage (drawing, p. 79).

To cut the more gradual arcs of the bigger windows, we used a 1x3 to extend the pivot point of the circle-cutting jig across the shop. The 2x arcs were screwed to a plywood base or to the subfloor, depending on how big they were.

Ripping Strips

Once the bending forms were completed, the strip-cutting operation was next. The basic principle here is that you're taking a piece of straight casing with the molded profile you want and ripping it into narrow strips that you can bend around a form and glue back together. But if you were to do this by ripping a single piece of casing, the resulting molding would be narrower than the original because of the material lost to the saw kerf. Therefore, you have to make alternate cuts on two pieces of straight casing.

To ensure that the laminating strips were cut to a uniform width, we used thin pieces of pine as spacers resting against a preset table-saw fence. This enabled us to cut the casing incrementally without changing the position of the saw fence.

In our case, the sawblade, and hence the laminating strip, was roughly ⅛ in. wide. The spacers were cut so that each was twice the width of the table-saw blade. We cut our spacers the same length as the short auxiliary fence on my table saw. To keep them from slipping with the casing as it was being cut, we simply tacked a brad to the underside of each spacer, which hooked over the front edge of the saw table (drawing, facing page).

After the spacers were completed, we adjusted both pieces of casing (ripped a little off them) so that the finished width was an even number multiple of the spacers. Our casing had a rabbeted back band around the outside edge, so we were able to reduce slightly the outside edge of the casing without changing the profile. (The side casings were also adjusted in width to make them equal to the arched head piece.)

Next, we glued and clamped the back band to the casing. We also filled in the plowed relief on the back of the casing by gluing in thin material and jointing it flush. This was necessary so that each strip would be cut square to the others.

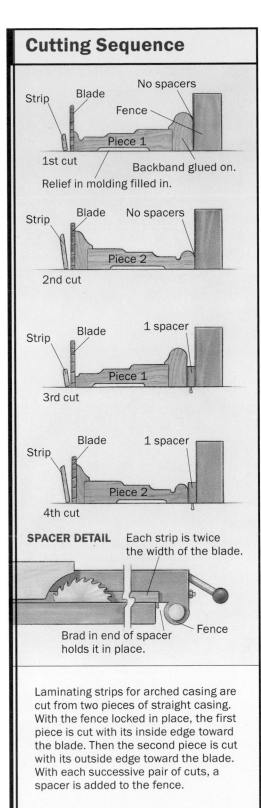

Cutting Sequence

1st cut
Strip · Blade · No spacers · Fence · Piece 1 · Backband glued on. Relief in molding filled in.

2nd cut
Strip · Blade · No spacers · Piece 2

3rd cut
Strip · Blade · 1 spacer · Piece 1

4th cut
Strip · Blade · 1 spacer · Piece 2

SPACER DETAIL — Each strip is twice the width of the blade.

Brad in end of spacer holds it in place. · Fence

Laminating strips for arched casing are cut from two pieces of straight casing. With the fence locked in place, the first piece is cut with its inside edge toward the blade. Then the second piece is cut with its outside edge toward the blade. With each successive pair of cuts, a spacer is added to the fence.

To ensure that the laminating strips were cut to a uniform width, Shafer kept the table-saw fence at a fixed distance from the blade and used spacer strips to move the straight stock incrementally closer to the blade. The rack beside the saw holds the laminating strips in their proper order for gluing.

We set the table-saw fence to equal the total adjusted casing width (casing plus back band) minus the width of one spacer. Finally, we equipped our table saw with a riving knife (or splitter) mounted behind the blade. This protected the thin strips from damage as they came off of the saw.

In order to produce alternating strips from two pieces of straight casings, we ripped the first piece of casing on the saw with the outside edge against the fence. Next, the second piece of casing was ripped with the inside edge against the fence, and so on.

On the first piece of casing that we did, I made the mistake of dropping the laminating strips into a pile, thinking I could easily sort them out later. Wrong. It took me more than an hour to put the pieces in order before gluing them up. For subsequent casing, I built a rack to hold the strips in order (photo, above). The rack was simply a pair of 1x4s with saw kerfs in them, nailed to a short bench.

Reusable Jig for Curved Jambs

With the trend toward New Classicism in architecture has come a renewed popularity of arch-top windows and, to a lesser extent, doors. Making the jambs for these doors and windows presents only minor challenges. The millwork is straight-forward, a matter of laminating a stack of plies in a curved jig. The problem is cost. One of the axioms in the millworking business is that no matter how many stock jigs you have cluttering up the back room, the next customer will want a window of a slightly different size. Building a custom jig for each order soon prices the work out of the market.

Over the last few years, David Marsaudon and the folks at San Juan Wood Design in Friday Harbor, Washington, have developed and refined a jig for making arch-top jambs that employs reusable parts. With this jig, they can lay up fair, smooth arcs of virtually any radius, and do so at a cost that gives them a competitive edge in the custom market. Their jig is made with some scrap plywood, strips of ¼-in. hardboard, a sheet of particleboard, and a shop full of clamps.

Building the Jig

The jig is made of three parts: a semicircular particleboard panel, a set of clamping blocks, and a mold surface. The radius of the particleboard panel determines the final size of the jamb.

I watched craftsmen Roger Paul and Jerry Mullis make the frame. They started with a given finished radius (the inside of the finished window frame) of 73 in. Based on previous trials, they estimated springback to be 2½ in., so the jig had to have a radius of 70½ in. By subtracting the thickness of the mold surface (¼ in.) and the depth of the clamping blocks (3 in.), they came up with a panel radius of 67¼ in. A new panel must be cut specifically for each size window frame, but smaller panels can be cut from larger ones to save on materials.

Paul and Mullis propped this panel vertically on a bench, and screwed to its rounded top a set of clamping blocks at about 8 in. o.c. Made from scraps of ¾-in. plywood laminated in pairs, these blocks are saved for reuse in each new jig. Finally, a skin of ¼-in. hardboard, as wide as the intended window jamb, was stapled to the clamping blocks. Paul and Mullis started at one end and carefully worked their way to the other, keeping the skin centered on the blocks and checking the final jig for bowing. Because any imperfections in the skin would telegraph through to the window jamb during glue-up, they checked that the skin was perfectly smooth. As a final touch, they waxed the skin with paraffin and smoothed the wax with steel wool. The wax keeps glue from sticking to the skin, so the finished jamb will pop easily out of the jig.

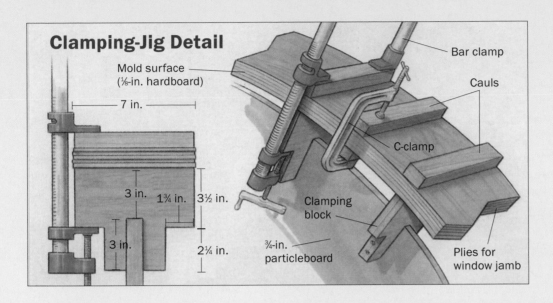

Clamping-Jig Detail

Mold surface (⅛-in. hardboard)

Bar clamp

Cauls

7 in.

C-clamp

3 in. 1¾ in. 3½ in.

Clamping block

3 in. 2¼ in. ¾-in. particleboard

Plies for window jamb

Assembling the Plies

Curved jambs are built from four plies, each ³⁄₁₆ in. thick, about ¼ in. wider than finished width to allow for edge-jointing later, and a few inches longer than arc length. Paul and Mullis first sorted among the four plies and selected the best for the inside finished surface. They finish-sanded this face—a job that is much easier on a flat bench before glue-up. They laid that ply face-down on the bench, and on its back they drizzled aliphatic glue. They spread the glue to an even film, paying special attention to coating the surface along the edges. Then, they lifted the second ply onto the first and repeated the procedure until all four plies were glued. They clamped the plies by the edges to keep them together as a unit until clamped in the jig.

Clamping

Working quickly before the glue set, Paul and Mullis lifted the group of plies onto the jig and balanced it on the apex. Starting at the center, they laid a caul across the width of the plies, centered over a clamping block. They hooked one bar clamp under the chin of the block, snugged it down, and then did the same with another clamp on the opposite side. Working as a team, one on each side of the jig, Paul and Mullis worked away from the center, clamping to each block with just enough pressure to hold everything in place. With the frame now held stable in the jig, Paul and Mullis went back and added two C-clamps, with a caul above and below, between each of the clamping blocks (drawing above). With all the clamps in place, they tightened them firmly, and glue flowed from every seam. They have found that only an even clamping pressure will give a fair arc. The key is to use lots of clamps and tighten them uniformly.

The next morning the clamps were removed, scraped free of dried glue, and hung back in their racks. With the bench clear, Paul and Mullis laid the finished frame against the penciled layout marks. The frame had sprung to within ⅛ in. of the design width.

J. Azevedo is a freelance technical writer in Friday Harbor, Washington.

The bending form is simply 2x stock cut into curved sections and screwed to a piece of plywood. Here the laminated strips have been clamped up without glue to work out the clamping strategy and eliminate some of the usual glue-up trauma.

Gluing Up

Before we could start gluing, the bending forms had to be adjusted to allow for the jamb reveal—the difference between the inside edge of the jamb and the inside edge of the casing. We tacked ¼-in.-thick spacing shims (the width of the reveal) against the

forms. We also covered the forms with waxed paper to prevent the casing from adhering to them during glue-up.

We had plenty of preadjusted clamps on hand to do the job—everything from bar clamps to wedges against wood blocks screwed to the floor. Our clamping cauls were bandsawn to match the outside radius of the casing. We dry-fit the strips around the form so that we could work out a clamping strategy (photo, facing page). During glue-up, we quickly and evenly brushed yellow glue on each piece. Because the casing was relatively wide and the setup time relatively short, we glued and clamped the strips in three stages and let the glue dry overnight before proceeding to the next stage. Once the complete casing was dry, we scraped and sanded the casing profile, removing glue squeeze-out and any irregularities in the profile.

Making Extension Jambs

We made the extension jambs for the windows with the same bending form used for the casing—all we had to do was remove the ¼-in. shims. The reveal was ¼ in. so we made the extension jambs ⅜ in. thick, allowing sufficient material to secure the casing. We produced strips roughly ⅛ in. thick to reduce the chance of springback and wide enough to fill the space between the window frame and the edge of the drywall, plus ¼ in. extra for ripping and jointing to the finished width after the glue had dried.

Ripping and jointing was a two-man operation. One man fed the piece into the table saw or jointer, and the other helped support the piece as it went into and came out of each machine. Cutting and fitting the extension jambs to proper length was a trial-and-error process, the error always being on the long side until the jambs fit. Next they were glued and nailed to the window frames through predrilled holes.

Fitting the Casings

The ease of fitting the casings to the windows was directly related to the care with which the pattern had been made. If the pattern was true to the window form, the casing was relatively true to the window.

Because our laminating strips were about ⅛ in. wide, the springback was negligible. We were using relatively wide casings, so springing the casing to match the window—anywhere the pattern was not true—was very difficult, if not impossible. We had to live with compromises in a few places.

Just as with the extension jambs, fitting the casing was a trial-and-error process. It was relatively simple on the windows with one-piece casings that were butted directly to the window stools. Likewise, using plinth blocks would have simplified fitting the casing on the bigger windows and doors. But we decided to miter the corners between the arched casing and the side casings.

We calculated the miter by tracing the head casing and side casings right on the drywall, then connecting the points where their inside edges and outside edges intersected. The head casing was cut to match this line, and the corresponding angle was then cut on the side pieces. After the miter was judged to be tight, the bottoms of the side pieces were marked and cut square to rest on the window stool, or mitered if the window was picture-framed.

Jonathan F. Shafer was a carpenter for almost 20 years. He is now the catalog director for a tool and machinery company.

The ease of fitting the casings to the windows was directly related to the care with which the pattern had been made.

Bench-Built Window Trim

■ BY JIM BRITTON

Back in the early 1980s, I saved a general contractor a bushel of money by pre-assembling a massive window-trim unit (for 12 windows), and then lifting it into place as a single component. I could do this because all the windows needed jamb extensions, which acted as reinforcement for the mitered casings for each window. So I built the trim on the floor near the windows, and a fellow worker helped me set the unit in place. With only minor tweaking, I was able to plumb and line the trim to near perfection. Since that day, I have never trimmed a window a piece at a time.

Preassembled Casings and Jambs Begin with the Stool

I'd say about 80% of the windows that I install need jamb extensions. Windows are typically made for 2x4 walls, and they end up in walls framed with 2x6s (photo 1). You can pay a premium for windows with jamb extensions, but the builders I know usually make their own to save a little money. I typ-

ically make them out of preprimed, ¾-in.-thick finger-jointed pine, the material used in the project illustrated here. But you can also use any of the other popular trim woods or medium-density fiberboard. The first step in making jamb extensions is to decide how wide to make them. I measure the distance from the window to the finished wall at the top and bottom of the window, and then split the difference. This works fine if the discrepancy is no more than ⅛ in. If it's more, you'll have to cut tapered jamb extensions, a sure way to bring a production pace to a virtual halt.

My first cut is the head-jamb extension. Its length is equal to the width of the window plus the reveals on both sides of the window. I chose a ³⁄₁₆-in. reveal for this job, so the length of the head-jamb extension equaled the width of the window plus ⅜ in. Next, I calculate the height of the side extensions. In this case, I measured the window and added ³⁄₁₆ in. for the top reveal, plus ¾ in. for the thickness of the head jamb. The length of the stool (photos 1, 2, 3) depends on the reveals, the width of the side casings,

Jam Extensions Reinforce Preassembled Window Casings

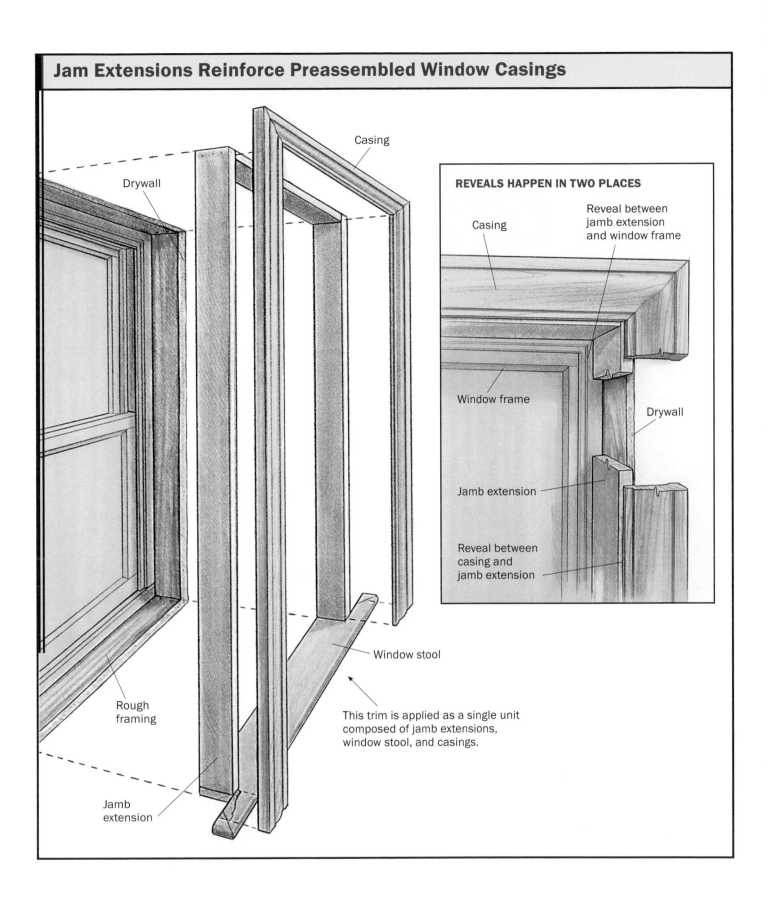

Drywall

Casing

REVEALS HAPPEN IN TWO PLACES

Casing

Reveal between jamb extension and window frame

Window frame

Drywall

Jamb extension

Reveal between casing and jamb extension

Rough framing

Jamb extension

Window stool

This trim is applied as a single unit composed of jamb extensions, window stool, and casings.

PREBUILT WINDOW CASINGS START AS A BOX COMPOSED OF JAMB EXTENSIONS AND THE WINDOW STOOL

1. Windows that are made for 2x4 walls end up in a lot of walls framed with 2x6s. The resulting raw edges between the window and the drywall will be covered by jamb extensions on the sides and top, and by the window stool at the bottom. Here, the author marks the stool for the notch that will create the horn, the portion of the stool that extends beyond the window casings. Note how the middle of the stool is registered against a line that marks the center of the window opening. **2.** Next, he marks the plane of the drywall on the end of the stool to show the cutline. **3.** The window-stool horns are then mitered for their returns. **4.** The side jambs are affixed to the stool with a pair of drywall screws at each corner.

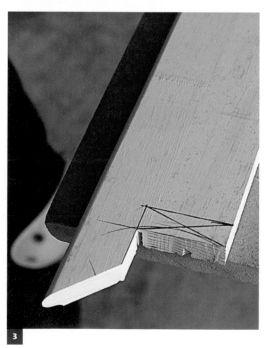

5. Britton marks the cut on the edge of a piece of casing. The cut mark allows for the width of the reveal between the edge of the jamb extension and the casing. **6.** Using a 16-ga. nailer, the author next attaches the casings to the jamb extensions. At this stage, resist the temptation to nail through the mitered corners or the bottom of the stool to the casing. To do so eliminates some of the flexibility of the assembly, which will work to your favor during installation. **7.** Before installation, the author uses hot-melt glue to attach the mitered returns to the stool's horns.

and the distance that the stool projects beyond the casing—¾ in. for this job.

Once I've got the jamb extensions and the stool screwed together into a big, shallow rectangular box, I'm ready to attach the casings and the return miters at the ends of the stool (photos 5, 6, 7).

Jockey the Assembled Frame into Its Opening

I lift the trim unit into position and wiggle it around until I like its location with respect to the desired reveals. Then I fasten the assembly at the lower corners by nailing through the side casings into the trimmers (photo 8). Next, I fasten the top corners by nailing through the side casings near the top. The assembly is now located and ready for lining the head, sides, and stool. To line, simply move the casing, and the jamb will follow. When it's straight, nail through the casing (photo 9). I end up putting my nails about 12 in. apart.

To ensure that the jamb extensions butt into the window frame, I push on the extensions near the corners and nail them into the trimmer (photo 10). I'm not worried about the lack of backing here because the screwed connection between the jamb extensions keeps them from being driven out of alignment.

When I drive nails into the jamb extensions away from the corners, I always place them near the nails in the casings (photo 11). This placement keeps the nails from pulling the extensions too far inward. A note of caution here: This shimless system works just fine if the gap between the trim and the framing is no more than ⅜ in. If it's any more than that, I nail furring strips to the inside of the rough opening.

THE CASING BECOMES A NAILING FLANGE DURING INSTALLATION

8. Once the bench-built trim is in its opening, Britton tweaks its placement until the reveals around the edge of the window look right. Then he affixes it to the wall with nails at each corner, beginning at the bottoms. **9.** When the corners are fixed, Britton next adjusts the alignment of the jamb extensions by pushing or pulling on the middle of the casing. **10.** Casings affixed, the jamb extensions are pushed against the window frame and nailed. **11.** Nails into the jamb extensions should be placed near nails into the casings. If irregularities in the wall or window position cause the miters to remain open, use a shim to close them, and then drive a single nail across the corner to pin the casings together.

12

Finish with the Stool and Its Apron

Next, I nail home the stool, aligning it by eye with the bottom of the window as I support the stool with a pry bar (photo 12). The stool is now pinned in place and can be fine-tuned up or down. I use the steel bar to pry or tap it in line if necessary. If you think that you need solid backing under the stool, now is the time to add it. But truth be told, the nails will support the stool just fine unless somebody decides to stand on it. And remember, the apron also works to support the stool from below.

I prefer mitered returns on the apron (photos 13, 14). I cut these delicate pieces with a power miter box. Then I use hot-melt glue to affix them to the apron. I nail the apron to the framing under the stool with nails on 12-in. centers (photo 15). Like the rest of the assembly, the nails should be long enough to get at least ¾-in. penetration into the framing.

Jim Britton is a contractor and trim carpenter who lives in Jacksonville, Oregon.

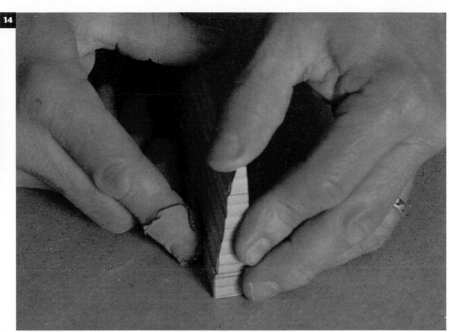

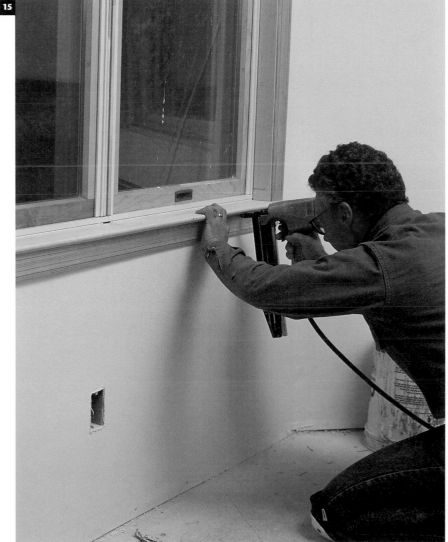

FINISH UP WITH THE STOOL AND THE APRON
12. With a pry bar holding the stool in line with the window, Britton nails the stool to the framing. **13, 14.** Mitered returns glued to the ends of the apron make for a delicate detail that eliminates end grain. **15.** The apron conceals any gaps between the stool and the drywall while simultaneously supporting the stool.

Crown Molding Basics

■ BY TOM LAW

The first piece of molding (right) is cut square and run into the corner. The second piece (left) is cut to the shape of the molding's profile (coped) and will butt neatly into the face of the other piece. The paper-thin point on the bottom of the coped piece will make the finished joint look like a miter.

The old-time carpenters I learned from used to amuse themselves by quizzing young apprentices about the trade. If you could answer the easy questions, the last question would always be: "How do you cut crown molding?" And when you looked puzzled, they'd go off, chuckling to themselves something about "upside down and backwards." Of all the different moldings, crown molding is the most difficult to install,

largely because of how confusing it can be to cope an inside-corner joint.

In classical architecture, crown molding (sometimes called cornice molding) is the uppermost element in the cornice, literally crowning the frieze and architrave. These moldings were functional parts of the building exterior when the ancient Greeks used them, but they have been used for centuries on interiors purely as decoration.

Crown molding is installed at the intersection of the wall and ceiling. Originally crown molding was triangular in cross section—the portions abutting the wall and ceiling formed two sides of a right triangle, and the molded face was the hypotenuse. But only the molded face is visible, so much of the solid back has been eliminated to save material. Also, by eliminating part of the back, only two small portions of the molding bear on the wall and ceiling surfaces, which makes crown easier to fit to walls and ceilings that aren't straight or that don't form perfect right angles.

Measuring and Marking

Crown molding can be used by itself or combined with other moldings, but it should always be in proportion to the size and height of the room in which it's installed. Too much molding at the ceiling line tends to lower the ceiling visually. Three or four inches of molding at the ceiling line is about right for an average-size room.

In the rooms shown in this article, I used 3⅝-in. crown molding, which is the most common size available. This dimension is the total width of the molding, but it's not the critical dimension that you use when installing crown. You need to know the distance from the intersection of the wall and ceiling to the front of the molding, measured along the ceiling. Because the back part of the molding has been eliminated, you can't measure this directly. Instead, I put the molding inside a framing square to form a triangle and read the distance (drawing, right). The molding shown here measures 2⅟₁₆ in. I mark that distance on the ceiling at each corner of the room and in several places along the walls. These marks will serve as a guide when I install the molding.

It's frustrating to drive a nail through a piece of molding and not hit anything more

solid than drywall, so I locate the framing members ahead of time—when I can still make probe holes in the wall that will be hidden by the molding. If the room hasn't been painted, you can spot the studs and ceiling joists from the lines of joint compound and make pencil marks on the wall to guide the nailing. If the room has been painted, you can find the studs and joists by tapping with a hammer and testing with a nail. Electrical outlets and switches are nailed into the sides of studs and offer a clue to stud locations.

Running crown molding should be one of the final jobs on a new house. Walls and molding should be primed and first-coated, the molding installed, and then finish coats applied. If you're retrofitting crown molding, it should be prefinished entirely so that all you need to do is touch up the paint or stain after installation.

On this job the walls were painted and the molding was prestained, so I marked the stud locations right on the crown molding. Rather than use the pencil to mark the wood, I made a slight hole with the point of a nail, which was easier to find in the dark stain and which I later nailed through.

TIP

If you're retrofitting crown molding, it should be prefinished entirely so that all you need to do is touch up the paint or stain after installation.

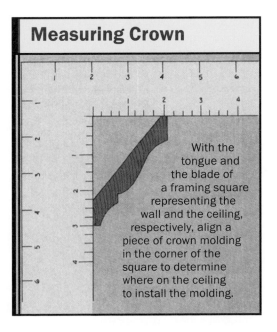

Measuring Crown

With the tongue and the blade of a framing square representing the wall and the ceiling, respectively, align a piece of crown molding in the corner of the square to determine where on the ceiling to install the molding.

The Crown Tools

When I cut crown, I like to work right in the room where the molding will go so I can orient myself to the wall I'm working on. If a room is finished, however, I may have to do the cutting somewhere else. Then I have to imagine the molding in place when I'm positioning it in the miter box (and believe me, this can get tricky with crown molding).

I cut and install crown molding with hand tools. I use a wood miter box because it's the kind I learned on, but also because my view is not obstructed by the electric motor of a power miter box. Installing crown molding is slow and calls for careful work, so the production speed of an electric miter box is not required. I cut miters with a standard 26-in. handsaw (10 or 11 point). Miter cuts are made through the face of the molding, and a sharp handsaw will do a better job than a dull circular-saw blade will do.

For this kind of work, I prefer a workbench to a sawhorse. Mine is just a simple frame of 2x4s and 1x4s with a 2x12 top. It

stands 34 in. high, which is a more convenient height to work on than a sawhorse provides. You don't need to deliver a lot of power to cut trim. A broad benchtop is also convenient for holding tools.

Although you can still find deep-throated coping saws in the mail-order catalogs, most coping saws nowadays are 5 in. deep and have a 6-in. blade. The blades come with different numbers of teeth. I try each kind of blade to see which works best with the wood I'm cutting. Generally, finer teeth work best with hardwood and coarse teeth with softwood, but not always. For the job shown here, I used a fine-tooth blade to cut softwood. It works more slowly, but makes a smooth cut.

The blade of a coping saw can be inserted with the teeth directed toward you to cut on the push stroke. Although it's strictly a matter of personal preference, I orient mine to cut on the push stroke because it acts in the same manner as a handsaw.

Before the crown molding can be coped, the end must be mitered to expose the profile. Positioned "upside down and backwards" in the miter box, the molding rests against small nails that hold it at the proper angle.

Getting Started

When I run crown molding in a typical room—four walls, no outside corners—I usually start with the wall opposite the door (drawing, below). Unless it's perfect, a coped joint looks better from one side (looking toward the piece that was butted) than it does from the other (looking toward the piece that was coped). By first installing the crown molding on the wall opposite the door, and coping the molding into it on both ends, the two visible joints show their best side to anyone entering the room.

I put the first piece up full length with square cuts on both ends. I cut it for a close fit, but if it's a little short I don't worry. Any small gap will be covered by the coped end of the intersecting piece. I hold the molding in place, lining up the top edge on the 2⅛-in. mark, and nail it. I use the shortest finish nails that will reach the framing, usually 6d or 8d. I nail into the wall studs through the flat section of the molding near the bottom, and I nail into the ceiling joists or blocking through the end of the curve near the top. I don't nail too close to the ends when I first put the piece up. I leave them loose to allow for a little alignment with the intersecting piece.

Occasionally I need to pull the top edge of the crown tight against the ceiling, but there's no ceiling joist or blocking to nail into. When that happens, I use 16d finish nails to reach the double top plate on the wall. Or sometimes I put a little glue on the molding and drive a pair of 6d finish nails at converging angles into the drywall about ½ in. apart. This pins the molding to the drywall while the glue dries. Another trick for nailing up crown when you won't have adequate framing is to nail up triangular blocks as shown in the drawing on p. 101.

I usually work around the room from right to left because I'm right-handed, and making coped joints on the right is a little easier than on the left. The second piece of crown to go up needs to be coped on the right and square cut on the left.

Nailing Tip

When there's no joist or blocking to nail into, you can put some glue behind the molding, then drive a pair of 6d nails at converging angles through the molding into the drywall. This will hold the molding while the glue cures. The drawing also shows where to place the nails when installing crown.

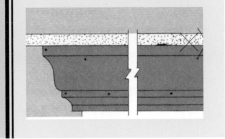

Molding Layout

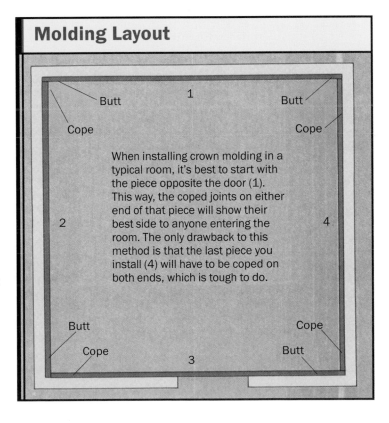

Butt   1   Butt

Cope   Cope

When installing crown molding in a typical room, it's best to start with the piece opposite the door (1). This way, the coped joints on either end of that piece will show their best side to anyone entering the room. The only drawback to this method is that the last piece you install (4) will have to be coped on both ends, which is tough to do.

2   4

Butt   Cope

Cope   Butt

3

After exposing the profile of the crown molding, the back part of the stock is cut away with a coping saw, which must be held at a severe angle or the coped joint will not be tight.

Coping with Crown

Finish carpentry can be harder at times than cabinet work because you often have to make perfect joints against imperfect surfaces. Coping inside joints rather than mitering them is one way to deal with that problem. If you miter the inside corner with crown molding, the joint will often open up when you nail the piece because the wall gives a little. A coped joint on crown molding won't open up and will be tight even when the walls are not exactly 90° to each other.

A coped joint is like a butt joint, with one piece cut to fit the profile of the other. The first piece of molding is cut square and run into the corner. The second, or coped piece, is made by cutting a compound miter on the end to expose the profile of the molding, then sawing away the back part of the stock with a coping saw, leaving only the profile. The end of the coped piece will then butt neatly into the face of the first piece.

Because I don't always get the cope right the first time, I start with a piece of molding longer than I need and cope the end before cutting it to length. The phrase "upside down and backwards" refers to the position of the crown molding in the miter box when you're coping it. Crown molding isn't laid flat against the side or bottom of the miter box; it's propped at an angle between the two, just as it will be when installed. But the edge that will go against the ceiling is placed on the bottom of the miter box and is therefore "upside down." The right-hand side (if that's the coped end) is placed on the left and is "backwards."

The crown has to be positioned so that the narrow flat sections on the back of the molding, which will bear against the wall and ceiling, are square against the side and bottom of the miter box. When they are, the bottom should measure out the required 2⅛₆ in. Once I find this position, I usually draw a pencil line on the bottom of the miter box to help me position subsequent pieces. Sometimes I'll even put a few nails on the line or glue a strip of wood to it.

Even with the molding positioned correctly in the miter box, it's still easy to cut it wrong. When I make the 45° cut to expose the profile of the molding for the cope, I remind myself that I want to cut the piece so the end grain will be visible to me as I look at it in the miter box.

Coped joints are always undercut slightly, but crown molding has to be heavily undercut through the S-curve portion of the crown (called the cyma recta) or it will not fit right. I start the cope at the top of the molding. Having trouble going from the straight cut to the curve, I back the saw out and come in at a different angle to cut away the waste. I begin the curved line with a heavy undercut and hold this angle all way through. I cut as close to the profile line as I can (photo, above).

The bottom of the crown molding is made up of a horizontal flat section, a cove, and a vertical flat section. I cut down to the upper flat and then take the saw out and start cutting from the bottom. Some carpenters simply square off the bottom, but I try to leave the little triangular piece intact. I support it with my thumb as I'm coping and slice it paper thin. This little piece makes the coped joint look like a miter and helps close

Even with the coping saw cutting at a severe angle, it's tough to remove enough wood through the S-curve in crown molding. Additional stock often has to be pared away with a utility knife.

any small gap if the first piece didn't fit tightly to the wall.

I always test the cope against a scrap piece of molding to make sure I'm in the ballpark before actually trying it in place. Despite my best efforts to undercut the curved section, I usually have to pare away some more wood with my utility knife (photo, above).

I cut the piece just a little long and test it in place before cutting it to final length. If the fit of the coped joint is close, but still a little off, I can sometimes improve the fit by twisting both pieces either up or down the wall at this point—the 2⅛6-in. mark on the ceiling isn't sacred. The buildup of spackle or plaster in corners can distort the intersection of wall and ceiling. Some carpenters carry a small half-round file with them to fine-tune the fit of the cope.

Around the Room

Once the coped joint fits, it's time to cut the piece to length. You can measure the total distance from wall to wall, but I find it easier to measure from either of the two vertical flat sections on the molding that the coped piece will butt into. If I'm working alone, I either step off the measurement with a measuring stick (a 12-ft. ripping, for instance), or

I'll drive a nail into the wall (above the line of the crown molding) and hook the end of my tape measure over it. Wherever I measure from the wall, I'm careful to measure to the same place on the piece I'm cutting.

When the coped piece is cut to length, I nail it up just like the first piece, leaving the square-cut end unnailed for the time being. If I need to draw the coped joint tighter, I nail through the coped piece into the piece it abuts.

The third piece of crown molding goes up just like the second, but the fourth one needs to be coped on both ends (drawing, p. 97), assuming the wall is short enough to be covered with a single piece of molding. I cut this piece about ⅛6 in. longer than the actual measurement, bow out the middle, fit the ends, and snap it into place. The extra length helps to close the joints.

Some carpenters don't like having to cope the last piece on both ends because there's very little margin for error. The way to avoid this goes all the way back to the first piece of crown molding that's installed. Rather than put up the first piece with square cuts on both ends, you can temporarily nail up a short piece of crown molding and cope the short piece into it (top photo, p. 100). Then take down the short piece, work on

The phrase "upside down and backwards" refers to the position of the crown molding in the miter box when you're coping it.

If you install the first piece of crown molding in a room by cutting both ends square, the last piece will have to be coped on both ends. To avoid this, you can put up a short piece temporarily and cope the first piece of crown molding into it.

around the room, and slip the butt end of the last piece behind the first cope that you made. This way all four pieces of crown molding in the room will have one square-cut end and one coped end.

When I go into a room that's not a simple rectangle, the decision about where to start is influenced by where I'll end. If there is an outside corner in the room, I like to end by installing the shortest piece that has an outside miter. That way, there's less wood wasted if I cut it too short. If there's not an outside corner, I like to work so that the last piece is installed on the longest wall that can still be done with a single length of molding.

When I need more than one piece to reach from corner to corner, I cut the moldings square and simply butt them together rather than use scarf joints or bevel joints. Butt joints are easier to make for one thing. And for another, although wood isn't supposed to shrink in length, the truth is it does. Over the years, I've seen a lot of joints that have opened up, and of those, the butt joints looked better than the others.

Outside Corners

These are also mitered with the molding upside down and backwards in the miter box, but the saw is angled to bevel the piece in the opposite direction. When you miter for a cope, you expose the molding's end grain, but with a mitered outside corner, the end grain is behind the finished edge. Sometimes I cut them at an angle slightly greater than 45° to ensure that the outside edges mate perfectly. I usually add a little white glue, then nail through the miter, top, and bottom from both sides.

Sometimes outside corners will close tightly but the leading edge of one piece overhangs the other, perhaps because the corner is not exactly 90° or because one piece of molding is thicker than the other (more about that in a minute). If the molding hasn't been painted or stained, I'll trim the overhanging edge with a sharp chisel and sand it. This actually leaves a narrow line of end grain exposed at the outside corner, but once the molding is stained or painted, the end grain isn't very obtrusive. There are times when the molding has the finish coat already on it, and I can't do this because it would expose raw wood. In that case, I use my nail set to burnish the projection smooth (photo below).

On this house, I ran crown molding in the foyer and had to terminate the molding at the stairwell opening. I ran the molding

If an outside miter is open just slightly, sometimes you can close it by burnishing the corner with a nail set.

When a line of crown molding has to be neatly terminated on an open wall, the end should be mitered and "returned" into the wall with a small piece of molding. To avoid splitting such a delicate piece, it's best simply to glue it in place.

through the dining room, turned the corner at the stair, and ended the molding with a return—a mitered piece that caps the end of the molding. To make a return, I simply cut a miter for an outside corner on the end of a scrap of molding, then lay the piece face down on the bottom of the miter box and cut off the end. I glue this in place with white glue so as not to take a chance on splitting it by using a nail or brad (photo above).

What Can Go Wrong?

Whether because the wood was wet when it was milled, or because the knives were dull, or because of internal stresses in the wood, the exact dimensions and profiles of the pieces in a given bundle of stock molding vary considerably. The differences aren't obvious until you try to fit an inside or an outside corner with two pieces that don't match. It's best to make joints from the same piece whenever possible.

There are times when the wall or ceiling is so crooked that gaps are left along the length of the crown. If there is a short hump that causes gaps on each side, I scribe the molding and plane it for a better fit. If the gaps aren't too bad, it may be best to fill them with caulk. Another trick I've used is that of leaving a small space (usually about

Nailing Tip

When nailing crown molding, walls parallel with the ceiling joists are often a problem. A length of 2x4, cut to fill the space behind the crown (slightly smaller) and spiked into place, will provide good nailing.
– S.A. Inserra, Jamestown, NY

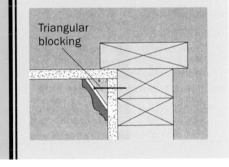

Triangular blocking

¼ in.) between the top of the molding and the ceiling, which makes it harder for the eye to pick up irregularities. If I'm doing this, I put up blocks to nail to, as shown in the drawing above, and use a ¼-in. spacer block to ensure a uniform reveal.

Tom Law is a former consulting editor for Fine Homebuilding. *He lives in Smithsburg, Maryland.*

Cutting Crown Molding

■ BY STEPHEN NUDING

few years ago, I purchased an 8½-in. compound-miter saw. It was light and compact, but had the same capacity for cutting large crown moldings as a regular 10-in. miter saw. Remodeling Victorian homes, I install a lot of crown so this seemed to be the perfect power tool for me.

I eagerly brought the saw to the job and set the miter and bevel angles for 90° corners, as indicated by the instruction manual. When I cut my first lengths of crown, the joints weren't perfect, but I figured that the walls and ceiling weren't perfect either, so with a little shaving here and there I was in business.

The next crown molding I had to install, however, was a larger one, and when I cut it and held it up to the ceiling, I was looking at a pie-shaped gap ⅜ in. wide. What's more, this room had two corners that were 135°, not 90°, and the saw's instruction manual gave no miter or bevel angles for this situation. I soon discovered that throwing miscut pieces around the room in rage and frustration is a very slow and expensive way to complete a job.

By now I was ready to return the saw to the dealer and demand a refund. But in desperation I grabbed the instruction manual one last time. According to the manual, the miter- and bevel-angle settings were correct for 90° corners when using a standard crown, which makes a 38° angle to the wall. Wait a minute—what if my crown doesn't make a 38° angle to the wall?

Fortunately, my daughter's protractor was in the car, so I was able to measure the angle the crown made to the wall by holding it against the inside of a framing square. The angle was more like 43° or 44°. I checked all the crowns I was installing only to find that none were the same, varying from 35° to 45°.

I finished the day's work as best I could and went home determined to calculate the angle settings for each of the crowns. Using my wife's high school math text to brush up on some trigonometry, I wrote down equations and measurements. I worked late into the night, but couldn't come up with a formula.

I finished the crown job eventually by trial and error, playing with the angles on the saw until they were right. Still, the problem gnawed at me. I spent a lot of late nights scribbling and thinking, but I just couldn't get it.

Fortunately, I had hung some French doors in the home of Roger Pinkham,

To miter crown with a standard miter saw, turn the molding upside down and set it at an angle between the fence and table.

To miter crown molding with a compound-miter saw, lay the molding flat on the saw's table.

professor of mathematics at the Stevens Institute of Technology®. So one Saturday morning, at my request, he graciously came to the house and we pored over my notes. Several hours later, we had it. We could calculate the miter and bevel angles for any crown and for any angle.

So, Why Use a Compound-Miter Saw?

You are probably wondering why anyone would want to calculate angle settings for a compound-miter cut when crown molding can easily be cut on a regular miter saw with no math at all. With a regular miter saw, the crown is positioned at an angle between the fence and the table (photo, top left), but is turned upside down so that the wall face of the crown is against the saw's fence and the ceiling face of the crown lies on the saw's table. The crown is cut at 45° to create a 90° corner, 22.5° for a 45° corner, and so on.

Most 10-in. miter saws, however, can only cut crown molding up to about 4½ in.

wide. Five and one half-inch crowns are readily available, though, and cutting these requires a 14-in. or 15-in. miter saw—a large, heavy tool. Cutting large crowns on any of these saws also requires the extra step of constructing a jig or fence extension, preferably both. Even a 15-in. miter saw is not big enough to cut crown molding more than 6½-in. wide, and larger crowns are also available. For instance, the Empire Molding Co., Inc. makes 8¾-in. crown that I often use.

So unless you want to make a king-size miter box and cut the molding with a handsaw, you'll have to use one of the new slide compound-miter saws or a radial-arm saw to cut these wide crown moldings. With a compound-miter saw, crown molding is laid flat on the saw table (photo, top right). No jig or fence extension is necessary. The saws can be smaller for cutting the same size crown, resulting in a lighter tool with a smaller blade, which is therefore cheaper to buy and costs less to sharpen.

Throwing miscut pieces around the room in rage and frustration is a very slow and expensive way to complete a job.

Crown molding varies not only in size but also in the angle that it makes with the wall. So the first step in calculating miter and bevel angles is to measure the crown with a framing square and determine the measurements shown in the drawing. Then plug those figures into the formulas shown.

BEVEL ANGLES

$$90° \text{ corner: } B = \sin^{-1}\left(\frac{D}{\sqrt{2} \times C}\right)$$

$$\text{Odd-angle corner: } B = \sin^{-1}\left(\frac{D \times \cos(F \div 2)}{C}\right)$$

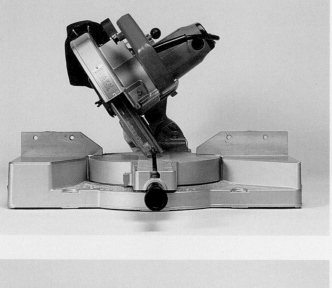

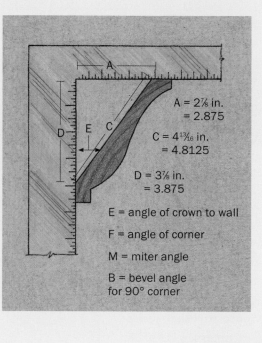

A = 2⅞ in.
= 2.875

C = 4¹³⁄₁₆ in.
= 4.8125

D = 3⅞ in.
= 3.875

E = angle of crown to wall

F = angle of corner

M = miter angle

B = bevel angle for 90° corner

MITER ANGLES

$$90° \text{ corner: } M = \tan^{-1}(A \div C)$$

$$\text{Odd-angle corner: } M = \tan^{-1}\left(\frac{A}{C \times \tan(F \div 2)}\right)$$

Figuring the Angles

To calculate the miter and bevel angles for any crown molding, you'll need a framing square and a calculator that's capable of doing trigonometric calculations. These calculators are usually called "scientific calculators." No cause for alarm, though, just think of yourself as carpentry scientist.

So here we go. First let's consider the most common case, the 90° corner. Hold whatever crown molding you're using up to the inside of a framing square as in the drawing above. Measure lines A, D, and C to the nearest 16th of an inch. (To convert fractions of an inch to decimals, simply divide the denominator into the numerator. To convert ⅞, for instance, divide 8 into 7 and you get .875.) The miter-table setting (M in our equation, is the inverse tangent of (A divided by C).

$$M = \tan^{-1}(A \div C)$$

To calculate this, divide A by C, and then hit the inverse tangent button (tan⁻¹), or arc tangent button (same thing). In our example, 2.875 (A) divided by 4.8125 (C) = .5974. With .5794 still on the calculator screen, hit the inverse tangent button and you get 30.9° (rounding to the nearest tenth of a degree). This is the miter angle at which to set your saw.

The bevel angle (B in our equation) is the inverse sine of D divided by (the square root of 2) times C.

$$B = \sin^{-1}\left(\frac{D}{\sqrt{2} \times C}\right)$$

To calculate this, multiply the square root of two (done on the calculator) times C. Then divide that into D and hit the inverse sine button, or arc sine (same thing) button on the calculator. Using the values from drawing A, the calculations would go like this: the square root of 2 = 1.41, times 4.8125 (C) equals 6.8059, divided into 3.875 (D) equals 0.5694, the inverse sine of which is 34.7°. This is the bevel angle at which to set your saw for a 90° corner.

Once you have calculated the miter and bevel angles for a particular molding, you never have to calculate them again as long as you have 90° corners. Jot down the angles somewhere and save a couple of minutes the next time you run that crown.

What if you have a wall corner that is not 90°? To make this calculation you'll need a device for measuring the angle of the wall corner. I use the Angle Devisor manufactured by Leichtung Workshops. Whether you are installing inside corners or outside corners, be sure to use the angle of the inside corner (the angle less than 180°) for the equation.

Here's how the equation looks:

$$M = \sin^{-1}\left(\frac{A}{C \times \tan(F \div 2)}\right)$$

If we were to use our crown from the drawing (A), we would have 135 (F) divided by 2 = 67.5. Hit the tangent button and you get 2.4142. That times 4.8125 (C) = 11.6184. Divide 11.6184 into 2.875 (A), then hit the inverse tangent button, and you get 13.9° (the miter angle).

For the bevel angle:

$$B = \sin^{-1}\left(\frac{D \times \cos(F \div 2)}{C}\right)$$

Plugging in some real numbers we get: 135 (F) divided by 2 = 67.5, the cosine of which is 0.3827. Multiply 0.3827 times 3.875 (D) and you get 1.4829. Divide that by 4.8125 (C), then hit the inverse sine button, and 17.9 appears. That's your bevel angle.

Finally, because the difference of one degree in the miter angle or bevel angle can be the difference between acceptable and unacceptable joints, you must set the angles on your compound-miter saw carefully. Math may be perfect, but measurements and the real world aren't, so slight adjustments may be needed to get an acceptable joint. But by using these equations you will avoid the fuss-and-fiddle approach I first used.

Stephen Nuding is a carpenter in Hoboken, New Jersey, who restores Victorian row houses.

Sources

Empire Molding Co., Inc.
721–733 Monroe St.
Hoboken, NJ 07030
(201) 659-3222

Leichtung Workshops
4944 Commerce Pkwy.
Cleveland, OH 44128
(800) 321-6840

Installing Two-Piece Crown

■ BY DALE F. MOSHER

I work as a finish carpenter on the San Francisco Peninsula, where there is a resurgent interest in formal houses that have a Renaissance European flavor. The houses often have a full complement of related molding profiles for base, casings, and crown, and to be in scale with the rest of the building, these profiles can be quite wide. In the case of the crowns, I'm talking 10 in. to 12 in. wide. In fact, the crown moldings that I sometimes install are so wide they come in two pieces.

There are several reasons for making crown in two sections. First, the machines that cut the moldings typically have an 8-in. maximum capacity. There is a lot of waste when wide moldings are carved out of a single piece of stock. For example, I'd need a 3x12 to mill a 10-in.-wide piece of crown—an expensive, inefficient use of the resource. Two-piece crowns are also a little more forgiving during installation. The type I used on the job shown here can be overlapped in and out a bit, allowing the width of the crown to grow and shrink as needed to account for dips and bows in the walls and ceilings.

All the two-piece crown moldings I've encountered have been custom-made. The designer or architect comes up with section drawings, and then the mill shop has the molding-cutter knives cut accordingly. Here, an average set of custom knives costs $35 per in., plus there is a $75 setup charge for each profile. So before the wood starts to pass over the cutters you've already spent a fair amount of money. But to create a certain look, it can be money well spent.

Stain-Grade or Paint-Grade Trim

If you've got one, your architect or designer will decide what grade the trim should be. If you don't, your checkbook will decide. At the mill, stain grade means the stock is clear and virtually free of knots. On the wall, stain grade means no opaque finishes will be applied. The moldings are individually scribe-fitted, and that means no caulks or putties to fill any gaps that may occur at miters and along uneven walls. Paint-grade material, on the other hand, will have some sapwood and uneven grain. On the wall, it can have caulkable gaps. Obviously, the stain-grade material will cost more—how much more depends on the species of wood. And it costs a lot more to install. Where I work, we figure four to five times more labor is needed to put up stain-grade crown as opposed to paint grade.

The most commonly used paint-grade materials in these parts are alder, poplar, and pine. I see more poplar than anything else because it's relatively inexpensive and easy to mill. Even paint-grade moldings, however, don't come cheap, and they should be handled with care. I have them primed on both sides as soon as they are delivered, and I store them on racks with supports no farther apart than 3 ft.

Backing Blocks

A two-piece crown needs a solid base for nailing and a flat surface to rest against to ensure correct alignment for the pieces. Backing blocks serve this purpose. To find the width of the backing block, I assemble a couple of short sections of crown, as they would appear when installed, and measure the backside from the point the assembly hits the ceiling to the point it engages the wall. The backing blocks should be ¹⁄₁₆ in. less than this measurement to ensure that the crown will go together without leaving gaps between the pieces, at the ceiling, or at the wall. Backing blocks can be made of solid wood, but I prefer ¾-in. plywood because it's affordable, doesn't split, and it holds nails well. After ripping a stack of backing-block stock, I cut the blanks into 6-in. to 8-in. lengths.

I prefer to place backing blocks on 16-in. centers, and no farther apart than 24 in. They should be affixed to the framing, so if the painting crew is about ready to prime the walls, I mark stud and joist locations with a keel (a carpenter's crayon) along the ceiling/wall intersection. A keel will bleed through most primers. Omission of this step gets you a one-way ticket to the planet of

Wide crowns, which appear to be made of a single molding, can be made by running related profiles adjacent to one another. Any gaps between the two are caulked prior to painting.

frustration, where you poke nails into the walls and ceilings, looking for the lumber.

Backing blocks are installed on layout lines snapped on the ceiling and wall. Using a torpedo level with a 45° bubble, I position a block at one end of the wall so the bubble reads level. The block should be about 2 ft. from the corner to avoid joint-compound buildup. I mark its edges on the ceiling and wall, and repeat the process at the opposite corner. These points are the registration marks for the chalkline. I use the raised chalkline as a straightedge to locate dips or bumps in the ceiling and walls. These problems are usually due to framing irregularities, and the backing blocks should be kept away from these places. I wish the framing crew could be around during this part of the job. If they only knew the trouble we go through to make them look good, they'd be taking trim carpenters to lunch a lot.

We figure four to five times more labor is needed to put up stain-grade crown as opposed to paint grade.

Miter-Box Station

I've used the new sliding compound-miter saws to cut crown, and I've decided to stick with my 15-in. Hitachi® chopsaw. Here's why: When you're running crown, you've got to make both back cuts and bevel cuts. It takes time to adjust the saw back and forth, and the constant changes multiply the chances for error. Also, the crown has to lie flat on the table with a slid-ing saw, which makes it harder to see the path of the blade and the cutline. None of these is a problem when using a chopsaw.

The key to cutting crown accurately is having a good station for the miter box. Mine has a pair of wing tables that flank the saw, connected by a ¾-in. MDF (medium density fiberboard) drop table that supports the saw. Each wing table is 6 ft. long and 2 ft. wide. They can easily be stood on end and carried through a standard doorway. I can also put a 2-ft.-deep table against a wall and have enough clearance behind the saw to swing it through its settings.

The wing tables have fences that support backer plates for the crown during a cut. To install them, I begin by snapping a line on the floor where the assembled station will sit. Then I put the outside legs of the tables on the line, and anchor each leg to the subfloor. Next I put the miter box in the drop section of the table. I removed the stock fences from the miter box, allowing me to bring the arbor slightly forward of the original fence line, thereby increasing the width of the cut. The wider of these two crowns was 7¼ in.—just about the limit of what my saw can handle.

I make sure the saw's turntable can swing freely from side to side, and that its blade is square with the tables when set at 0°. I secure the saw to the drop table with four drywall screws run through holes that I've drilled in the saw's base. Each wing table has an 8-in.-high fence that's square to the blade. I align the fences with a string to make sure they're straight.

During a cut, the crown bears against ¾-in. MDF backer plates that are screwed to the tables and the fences. The backer plates should be ½ in. wider than the widest crown section. The crown moldings

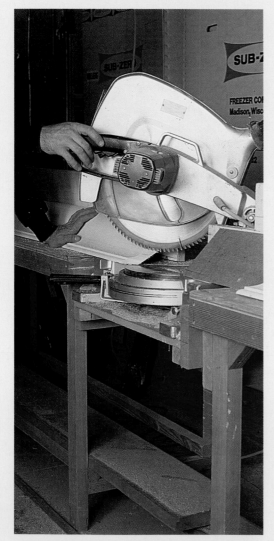

A pair of wing tables linked by a saw platform provide angled backer plates to support the crown moldings as they are cut. The plates are screwed to the table and fences.

The wing tables are joined by a saw platform lowered far enough to bring the saw's table flush with the wings. The fences are braced from behind with triangular blocks on 1-ft. centers.

for this job meet the wall and ceiling at 45°, so I ripped the backer-plate edges at 45°. Other crowns meet the wall and ceiling at different angles, and the backer-plate edges should be beveled accordingly.

It took me about a day to build this setup and another half day to fine-tune everything. But the time it takes to build one will be returned tenfold in a single good-sized installation job.

To attach the backing blocks, I use a 2¼-in. finish nailer because finish-nail heads are small enough to be consistently set below the face of the block. Nail heads that stand proud interfere with the crown. When I've got framing on one end of the block for nailing, but none on the other, I put a bead of glue on the backing block to help anchor it to the wall.

At inside corners, I run one block into the corner, and then scribe the adjoining one to it. The backing blocks are typically a little too wide to fit between the lines in the corners because of joint compound on the wall, and they need to be trimmed a bit to fit. Outside corners are sometimes mitered as though crown molding, and secured to the wall, ceiling, and to each other through their mitered edges.

When the backing blocks are up, I cut my "tester blocks" (photo, above). These are typically 16-in.-to 24-in.-long pieces of the crown molding. They need to be long enough to reach from an inside corner to the closest midspan backing block. I cut three pairs of tester blocks with inside miters. One set has 44° miters at each end, one has 45° miters, and the third has 46° miters. You

Crowns this wide need substantial backing to provide a consistent plane for aligning the two pieces and for adequate nailing. In the corner, one block extends to the wall while the other is scribe-fitted to it. Here, a test piece of crown is held in place. The pencil line along its point intersects the corner formed by the two blocks, marking the point from which the overall measurement for the crown will be taken.

may ask, "why not cope the inside corners?" For one, the curved profile of the widest molding in this job meant that a coped corner would be very fragile. I've found that a glued inside miter on a paint-grade job—if the pieces are carefully fitted—yields first-rate results.

Running Crown

Installing crown is not a solo operation. The job will go a lot faster and with greater accuracy if you've got a good helper. Working on the theory that a piece of trim can always be made shorter, we begin with the longest run in the room by tucking the pair of 45° test blocks into one of the corners, just the way the finished crown will fit. If the fit isn't

acceptable, we try a 44° and a 46° block until the right combination turns up. It might be a pair of 44s. It doesn't matter. It is very important, however, that the line of the miter lines up with the corner, whether it's an inside or outside miter.

Once we find the best fit, we make a pencil mark along the bottom of the block into the corner (photo, left). This marks the point from which the overall measurement is taken.

I don't bother to cut the piece a little long and then shorten it by degrees to ease into the fit. My helper and I can measure it accurately, so I cut it to that length. Period. This saves a lot of climbing up and down the A-frame scaffolds we typically erect as work platforms.

We also use tester blocks at outside miters to determine the best angles. To get our measuring points for an outside-to-outside miter, we make pencil marks on the ceiling to note the long points. For an outside-to-inside miter, we mark the long point of the outside miter, and the heel cut at the inside miter.

I'll typically put four 1½-in. finish nails into each backing block. The nails should be placed where the painter can easily putty the nail heads. I don't put nails in a tight radius or too close to an inside corner. A small prybar can be useful for aligning the crowns during nailing (bottom photo, facing page). I prefer the ones used by auto mechanics.

Back-beveling the miters can be useful on recalcitrant fits. A good tool for this is a 1⅛-in. belt sander. Its protruding belt makes it very maneuverable. If I need shims, I use pieces of manila folder. At each miter, I run a bead of yellow glue to ensure a sturdy joint.

Sometimes the crown has to work its way in short sections around a wing wall. In this case, I usually preassemble the pieces if they're shorter than about 12 in. I put the parts together with glue and a pneumatic brad nailer, let the glue set for 20 minutes, and then place it as a unit (top photo, facing page).

Short sections of crown are best preassembled into a single piece. The pencil marks on the ceiling show the points from which the crown-length measurements were taken. On the right you can see a fully assembled run of crown.

As we run the upper crown, we make notes in the corners that describe any special angles or back-beveling that it took to get a good fit. Nine times out of ten, the same cuts will work on the lower section of crown.

After the crown is up, the drywallers can float on any necessary topping compound to hide the bumps and bows in the ceiling and wall. If the walls are to be textured, this should be done after the crown is installed. Our painters use oil-based putty to fill the nail holes, and latex-based paintable caulk to make the joint between the two pieces of crown disappear.

Dale F. Mosher *is a carpenter who specializes in finish work in Palo Alto, California.*

As the moldings are nailed home, a small prybar is useful for aligning the adjoining sections.

Making Curved Crown Molding

■ BY JOHN LA TORRE, JR.

As a carpenter, I spend most of my time at work swinging a hammer or wielding a saw. But I'm always looking for a chance to try a new technique. Recently, while I was touring a house under construction, owner Paul Kreutzfeldt showed me a straight piece of stock crown molding he planned to use for the kitchen ceiling, then looked up at a curved wall in the room and said, "That piece is going to be a bear."

"Yup," I answered, excited. "Mind if I give it a try?"

"Go right ahead," he said with a smile.

There are two basic approaches to making curved trim. You can glue several pieces of wood end-to-end and then cut the curve on a bandsaw, or you can laminate thin strips of wood around a curved form. If you use the first method and decide to stain the trim, the separate pieces may accept stain differently, and the joints usually show through. Laminated trim, on the other hand, is stronger than butt-joined trim and usually looks better when it's stained. Because Kreutzfeldt had yet to choose between staining and painting his crown molding, I decided to laminate it. As it turns out, Kreutzfeldt decided to paint it.

Making the Bending Form

The first step was to make a form that matched the curvature of the convex wall. Finished with drywall, the wall defined a 90° arc having a radius of about 24½ in. Unfortunately, the curve was far from perfect, wandering out of round by up to ⅜ in. That forced me to make a template for the form.

To create the template, I bandsawed a 24½-in. radius curve in a sheet of ⅛-in. tempered Masonite. I then held this template against the curved wall 3¾ in. from the ceiling, which is where the bottom of the crown molding would contact the wall. After scribing the template with a pencil compass, I trimmed it with a jigsaw for a snug fit.

Making the bending form itself was the easy part. Back at my shop, I traced the outline of the template onto three pieces of ¾-in. plywood, cut them out, and nailed them together, placing ¾-in. plywood spacers between layers to produce the proper thickness. Finally, on the back edge of the form, I bandsawed a series of steps parallel to the front edge to give the clamps good purchase (photos, p. 114).

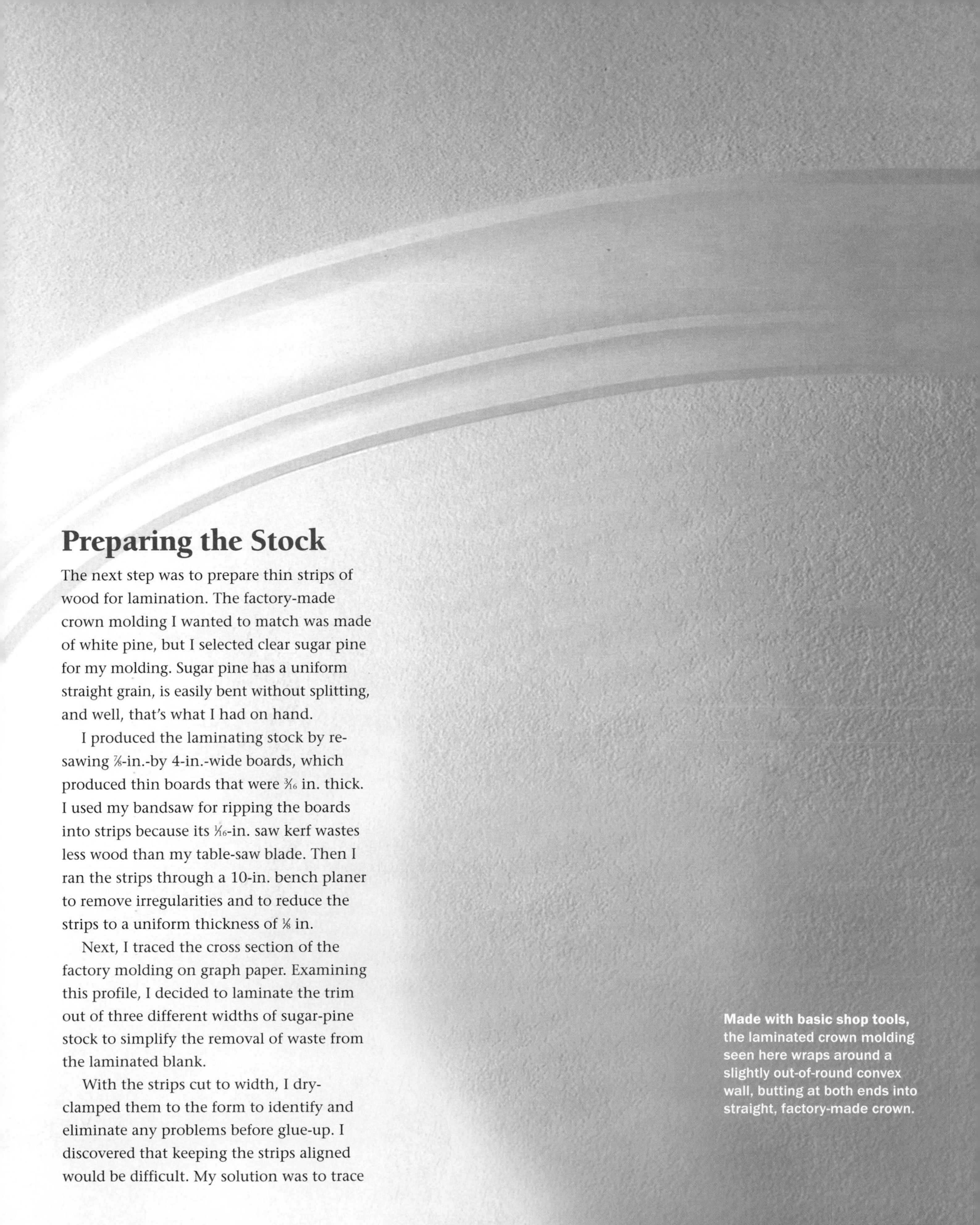

Preparing the Stock

The next step was to prepare thin strips of wood for lamination. The factory-made crown molding I wanted to match was made of white pine, but I selected clear sugar pine for my molding. Sugar pine has a uniform straight grain, is easily bent without splitting, and well, that's what I had on hand.

I produced the laminating stock by re-sawing ⅞-in.-by 4-in.-wide boards, which produced thin boards that were ³⁄₁₆ in. thick. I used my bandsaw for ripping the boards into strips because its ¹⁄₁₆-in. saw kerf wastes less wood than my table-saw blade. Then I ran the strips through a 10-in. bench planer to remove irregularities and to reduce the strips to a uniform thickness of ⅛ in.

Next, I traced the cross section of the factory molding on graph paper. Examining this profile, I decided to laminate the trim out of three different widths of sugar-pine stock to simplify the removal of waste from the laminated blank.

With the strips cut to width, I dry-clamped them to the form to identify and eliminate any problems before glue-up. I discovered that keeping the strips aligned would be difficult. My solution was to trace

Made with basic shop tools, the laminated crown molding seen here wraps around a slightly out-of-round convex wall, butting at both ends into straight, factory-made crown.

The laminating form consisted of three layers of ¾-in. plywood separated by ¾-in. plywood spacers.

A series of steps bandsawn into the back of the form provided solid footing for an assortment of clamps. Extra strips of pine placed against the molding stock helped distribute the clamping pressure.

a slightly oversized cross section of the three-step assembly on two scraps of plywood, cut the patterns out, and then slip the scraps over opposite ends of the assembly. These simple jigs helped to prevent the wood strips from sliding around during glue-up.

Shake It Up

For laminating jobs, I like to use urea-formaldehyde glue, which starts as a tan-colored powder that must be mixed with water. To make glue-up simpler, I taped the pine strips together edge-to-edge on my glue-up table. Then I spread the glue across the assembly using a paint roller. That done, I removed the masking tape, coated the masked areas with glue, tilted the strips upright, and pressed them together.

Glue-up took every clamp I had (including C-clamps, bar clamps, and pipe clamps), and it wasn't a pretty sight (photos, facing page). I installed the first clamp at the midpoint of the form and worked my way toward both ends, alternating clamps above and below the form. Extra strips of wood placed against the outer plies of sugar pine helped distribute the clamping pressure evenly. Excess was scraped off before the glue cured.

Glue Tip

I used to employ a stick or a rubber spatula for mixing glue but sometimes ended up with lumps of powder that wouldn't dissolve. Then I discovered a better method: Put the powder in a plastic container, add the correct amount of water, and then shake the container in a circular motion. Surprisingly, mixing in this way is faster than with a stick and produces a lump-free mixture every time.

After letting the glue cure for 24 hours, I removed the blank from the form. Checking the concave side of the blank against the template, I saw that the molding was within $\frac{1}{16}$ in. of a perfect fit. I decided not to shave it further just yet.

Sculpting on the Table Saw

Once the glue-up was completed, I ran the curved blank, top edge down, through the thickness planer to remove slight irregularities from the bottom edge of the blank. Then I flipped the blank over and planed its top edge to size. Planing the curved blank was easy—I simply steered it through the planer to keep it perpendicular to the cutterhead.

With the sizing completed, I squared both ends of the blank and traced the outline of the factory crown molding on one end. Now all I had to do was remove everything that didn't look like crown molding.

Probably the easiest way to make crown molding is to cut it on a shaper. Many cabinet shops nearby had a shaper, but none had a cutter that matched my molding. I could have ordered custom-made cutters, but that would have cost $300 to $400, difficult to justify for a one-off piece of trim.

I decided to shape the blank by making a series of table-saw cuts to remove most of the waste (photo, p. 116). This worked remarkably well. I clamped a $3\frac{3}{4}$-in.-tall board to the fence to make it the same height as the molding, then marked the top of the fence to index the centerline of the sawblade. While making each cut, I held the blank against the fence at the index mark. Any deviation from this mark was insignificant, because it merely caused the sawblade to wander into the waste area, requiring nothing more than a second pass across the table saw to get it right.

My blade cut a $\frac{1}{8}$-in. wide kerf, so I made the cuts by moving the fence toward the blade in $\frac{1}{16}$-in. increments, raising the blade

Laminated trim is stronger than butt-joined trim and usually looks better when it's stained.

The molding was contoured
by making a series of cuts
on a table saw to remove
the waste up to the layout
line. A mark on the rip fence
indexed the center of the
sawblade, indicating the
optimal spot for the molding
to contact the fence during
the cutting operation.

just enough each time to remove the maximum amount of stock without cutting across the layout line. In this fashion, I finished with a cross section very close to that of the factory crown.

Because the stock laid flat on the saw table, the cutting operation was accomplished safely and easily. I also kept my fingers far away from the sawblade at all times.

Scraping It Smooth

All that remained was to smooth out the small, sharp steps on the molding blank. I figured this would be the easy part, but it turned out to be the most difficult.

First I tried sanding, but the sandpaper quickly became clogged with pine resin. I soon realized I'd have to scrape the pine smooth. Dragging a homemade scraper along the molding at a 45° angle produced satisfying results (photo, facing page). Each

Make a Scraper

To make a scraper, I cut a 45° angle on the end of a scrap piece of the factory molding and traced its profile on the blade of an old taping knife. Then I cut the blade to the layout line using a bench grinder and raised a cutting edge by rubbing the blade with a hardened-steel punch.

pass left behind a small trail of fine dust instead of the curled shavings a perfectly tuned scraper would produce, but little by little the sharp steps began to disappear. Scraping the molding down to the layout line took two hours and a lot of elbow grease.

The scraping left the molding with some torn fibers and minor irregularities, so I decided to finish the job by sanding. I began by using 80-grit sandpaper to work out the irregularities, then worked my way up to finer-grit paper. I use Wetordry® TRI-M-ITE® sandpaper from 3M® Construction Markets Department, because its backing doesn't tear while sanding. The sanding took about three hours and just about wore me out.

Installation

Earlier, when trimming the ends of the curved blank, I had left the ends 1½ in. long. Now I used the bandsaw to cut a 1½-in.-long triangular stub tenon on both ends of the molding. These tenons would fit into the triangular voids behind the factory crown molding, aligning the joints while providing solid backing for the ends of the factory molding.

The final step was to fit and attach the curved crown to the wall. Using a belt sander, I relieved the concealed edges a bit so that the molding fit snugly against the wall and the ceiling. By now Kreutzfeldt had decided to paint the crown molding, but just a small amount of belt-sanding produced such a tight fit against the wall and the ceiling that no caulking was necessary. I applied construction adhesive along the back and top of the curved crown, then fastened the molding to the wall with screws run through the stubs.

As Kreutzfeldt installed the straight runs of crown molding, I was gratified to see that just a bit of sanding produced a satisfying match of curved to straight molding.

John La Torre, Jr., is a carpenter in Tuolumne, California, and a regular contributor to Fine Homebuilding.

Installing Crown Moldings

■ BY JOSEPH BEALS III

Crown molding has a royally painful reputation. Installation can be difficult: Unlike baseboard or casing, crown molding must sit at a consistent angle to the wall, making cutting and nailing more demanding. When the joints don't fit properly, when the nails hit nothing but air, and when the design that looked great on paper ends up looking trivial on the ceiling, the process of installing crown molding can be come extremely disagreeable. However, crown molding will yield to patience and to a few simple techniques that anticipate its frustrating behavior.

I've used some ceiling-trim designs repeatedly because they cover a range of stylistic options and because they're easy to build. These styles are not formal, but they go beyond the one-molding solution and add a surprising level of interest (drawing, facing page). The common element here is a piece of flat stock that I call backing trim. I usually shape a simple profile on the exposed edge; a scotia is shown in the photos, but ogees and beads are other possibilities. The backing trim can be as wide or as narrow as preference dictates. A narrow exposure will look like an additional crown-

Backing Trim Adds an Easy Detail to Crown Molding

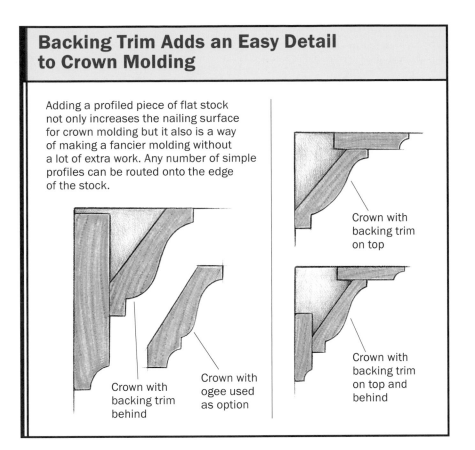

Adding a profiled piece of flat stock not only increases the nailing surface for crown molding but it also is a way of making a fancier molding without a lot of extra work. Any number of simple profiles can be routed onto the edge of the stock.

Crown with backing trim on top

Crown with backing trim on top and behind

Crown with backing trim behind

Crown with ogee used as option

molding element; a wide exposure becomes a design element in its own space. In the following pages, I'll describe the techniques I use to lay out, cut, and install this simple two-piece molding.

Use Chalklines as a Reference

After I've decided on the design of the crown mold, my next job is layout. For practically any design, I snap chalklines on the wall as a guide (photo, right). The layout lines should be straight between any two points, typically corner to corner, corner to window, between two windows, and so on. Because an out-of-flat wall is curved with respect to a straight line on the ceiling and vice versa, the layout lines will make this flaw visible, and the inevitable adjustment won't be unexpected. Snapped lines are also important for more formal designs that require blocking.

Because ceilings and walls are never flat and straight, the author prefers to snap lines that give him an accurate layout.

Designing Crown Molding

Known as the cornice, traditional ceiling-trim design echoes the entablature of a building, the place where the sidewall meets the roof overhang. In a classically influenced residential exterior, the entablature contains a frieze, a soffit, a fascia, and a cornice (inset drawing). At its most elaborate, the interior cornice's elements correspond to the exterior and include a crown, a fascia, a soffit, a bed, and a frieze. Most interior designs, however, are much simpler.

The most direct approach to trim design is to start at the top of the entablature and to pick as many elements as seem appropriate or desirable. The size of the room is not an important consideration: A large room with high ceilings requires elements of larger section, not more or fewer elements. Provided the parts are in appropriate scale, even a small room can support an elaborate cornice.

A model that's built from short sections of the proposed moldings is an ideal design tool that shows more detail than a flat drawing. It's easier to see the scale of moldings and embellishments in more complex designs; it's also a handy way to plan for blocking. Beware of designing too small, however. An old rule says that work to be placed overhead should be three times the size you think it should be. If a design looks good on a benchtop model, it will probably be undersize in place.

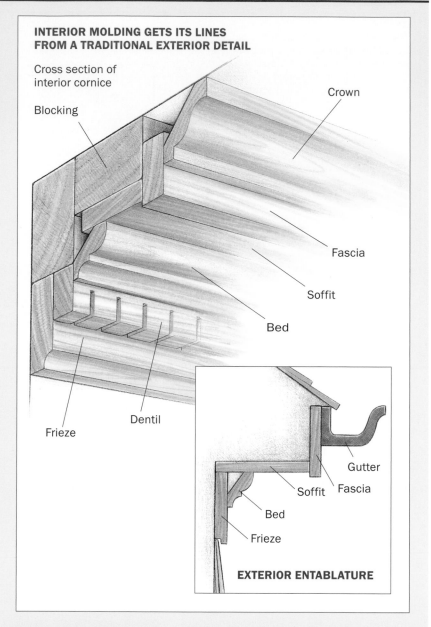

INTERIOR MOLDING GETS ITS LINES FROM A TRADITIONAL EXTERIOR DETAIL

Cross section of interior cornice

Blocking

Crown

Fascia

Soffit

Bed

Dentil

Frieze

Gutter

Fascia

Soffit

Bed

Frieze

EXTERIOR ENTABLATURE

If the room is really distorted, I like a big reveal in the design. The backing trim should follow the layout lines as much as possible, but the molding that comes after may be adjusted up or down a bit to conform to irregularities in the wall and on the ceiling. Remember that the trim serves the room, not a spirit level, so plan to make some compromises. A big reveal allows this to happen without appearing conspicuous.

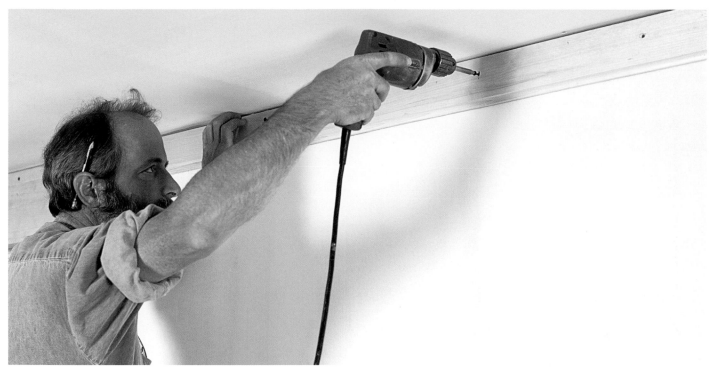
Screwed into the framing, a flat piece of trim alleviates the problem of nailing crown into irregularly spaced studs and also adds an extra level of detail.

Screws Hold the Backing

I like backing trim for its looks, but it also can solve serious nailing problems. Studding in old houses is often widely spaced, loose, or missing, usually where nailing is crucial. On the ceiling, strapping, lath, or joists can be hard to find or spongy. (During layout, I use a stud finder or drive a 6d finish nail into plaster to find solid nailing, and then mark the location with a pencil.)

I screw flat sections of backing trim in place because the holding power of a screw is huge, and there's no impact to distress old plaster. Also, a screw can get a solid bite in old wooden lath or a loose stud where nailing is useless. I like to hide the screws behind the molding, but if I can't do that, I counterbore first and glue in wooden plugs afterward. Even for painted work, the plugs will hide holes better than any other method of filling holes.

Reference Marks Keep the Molding Straight

Before I install the crown molding, I need reference marks on the backing trim. Snapped lines won't do the job here: I want the molding tight and accurately sprung, which means, as much as possible, maintaining the proper angle of the molding to the wall and ceiling throughout the length of any run. Sprung moldings touch the wall and the ceiling along two narrow surfaces. These surfaces are inadequate for ensuring a stable reference, and any molding, particularly large moldings, can shift. This shift creates frustrating problems where joints don't cope or miter properly. The displacement can happen when the molding wanders as it joins the wall to the ceiling, when it is pressed into a slow curve or rides past a hump, or when a twisted molding is pressed into a straight line. It can also happen with straight stock on a perfect surface because the last hammer blow shifted the molding. I

I like backing trim for its looks, but it also can solve serious nailing problems.

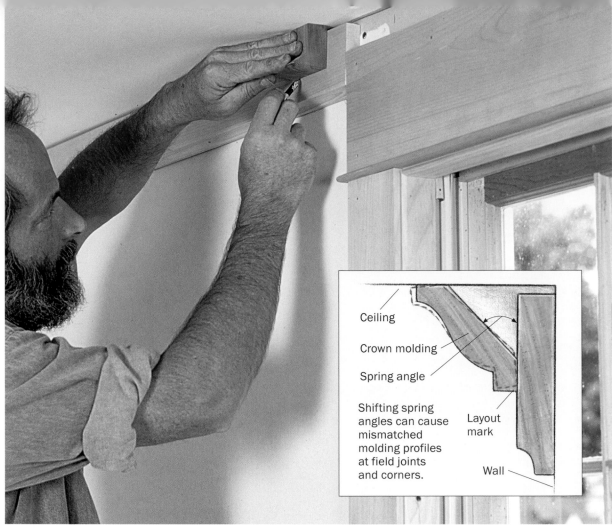

A gauge block shows exactly where the bottom of the molding must lie. Layout marks keep molding at a consistent angle to the wall, which in turn makes miters and field joints line up.

Ceiling

Crown molding

Spring angle

Shifting spring angles can cause mismatched molding profiles at field joints and corners.

Layout mark

Wall

guard against this condition by using a gauge block to mark the location of the molding's bottom edge.

To make the gauge block, I take several samples of the crown molding, hold each at the proper angle against a corner (such as the chopsaw table and fence), and mark the position of the bottom edge of the molding. The distance from the mark to the corner defines the height of the gauge block. A pencil mark ensures that I keep the molding to a constant spring angle and resolves many baffling fitting problems. I also use the same gauge-block method if I'm installing a sprung bed molding on a more complex cornice.

A Site-Built Jig Makes Cutting Crown easier

Once I've laid out the job, it's time to start cutting. The most confusing aspect of crown molding is that miters must be cut upside down. I also want to reproduce the angle of the molding between the wall and ceiling that I've just marked on the wall.

I make a jig for my miter saw's table that takes all the guesswork out of cutting. The jig consists of a fence that's screwed to a wooden auxiliary table (photos, facing page) that's bolted to the saw table. Registered against both fences, the molding is held securely and accurately during the cut. (For many people, it may be easier to build an auxiliary fence that's fastened to the saw's fence. Either way, the important thing is that you immobilize the molding's spring

With the gauge block used to mark the crown height on the backing trim, the author transfers a mark to a piece of masking tape applied to the fence. This mark corresponds to the mark on the wall measured by the gauge block.

A length of scrap wood keeps the molding's angle proper.

Holding a piece of molding upside down against the mark on the fence, the author transfers the molding's angle to the table.

angle as you make the cut.) Even if you have a compound-miter saw that can cut crown miters on the flat, it may take less guesswork to set up this way.

In any house, it's extremely rare to find a square outer corner. Rather than a fussy trial-and-error session with the miter saw, I've found that it's easier to mark the upper and lower points where the two pieces of molding meet and cut accurately to the lines.

To find the intersecting points, I hold the first piece of molding in place, trace a line along its upper edge on the ceiling, and mark the point where the bottom meets the corner (top photo, p. 124). I then repeat the process with the other side; now

On out-of-square corners, the author extends the upper line of the molding and transfers the intersecting points onto the stock.

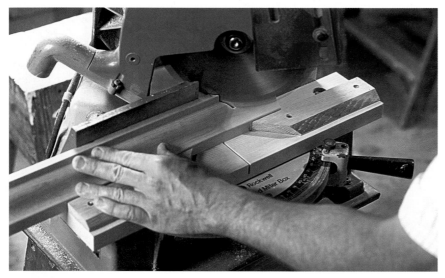

Aligning the sawblade to the marks, the author cuts an accurate miter with little guesswork.

The result is a tight-fitting intersection between the moldings.

I've drawn the upper part of the molding miter on the ceiling. After transferring the marks to the molding, I can cut an accurate miter (center photo).

On most jobs, I install the first run of the molding square cut at both of the ends, corner to corner. I then work counterclockwise around the room. I nail large moldings along both the top edges and the bottom edges, which is why backing trim can be so useful. I place the bottom edge on the reference marks, nail the bottom edge first, and then nail the top.

Coping Inside Miters Makes a Tighter Fit

Instead of mitering inside corners, I cope the right end of each successive piece of molding. A coped joint is essentially a butt joint made by scribing the end of one molding to fit the profile of the opposite half of the joint. Coping may take more time than mitering, but in the end, it makes a better joint. Often, inside miters will open as the joint is nailed because the molding is drawn tighter than it could be held by hand or because the drywall yields slightly under the influence of the hammer. The coped joint avoids these problems and can also be easily modified to accommodate small variations in the molding profile.

To make a coped joint, I run the first half of the coped joint tight to the corner. The adjoining piece is first mitered exactly as if it were half a conventional inside miter. If you hold a mitered molding in a corner close to an adjoining molding, you'll see how the miter cut reveals an edge profile of the adjoining piece. After darkening the profile edge with a pencil (top left photo, facing page), I use a coping saw to cut as close to the line as I can with a slight back cut (top right photo, facing page). If you look at the cut from straight on, you should not see any material protruding beyond the cutline. After sawing, I use a bastard file or a

Coped miters are labor intensive but they make tight joints. After cutting an inside miter, the author highlights the molding's profile with a pencil.

He cuts to the line with a sharp coping saw. Rather than make a square cut, he relieves the cut by a few degrees (called back-cutting).

The cut can be fine-tuned with a file to match the profile.

chisel to dress the cope until the profile fits (bottom photo).

Often, the last length of molding in a room must be coped at both ends. I first rough-cut the molding full length, cope the right side as usual, then measure the top and bottom lengths. I cut the miter in the chopsaw, sighting along the blade to make sure the angle and length are correct; cut the cope; and test the fit. A long molding is more easily fitted than a short one because

it can be cut a fraction oversize and sprung into place. A short molding must be perfect, and it can be tuned with a sharp chisel and rasp if it is almost right. Be prepared for mistakes, and dump a short molding if it just won't fit. A second attempt is a better investment of time than patching a bad joint.

Joseph Beals III is a designer and builder in Marshfield, Massachusetts.

Recycled Redwood Wainscoting

■ BY MILES KARPILOW

Over the 35 years that my wife and I have lived in our home, I have redone almost every room. Being a cabinetmaker, I started with the kitchen, then the bedrooms, putting back windows that had been covered over, replacing inappropriate ones, replacing cracked and loose plaster with drywall, and so on. When we bought this house, I knew it was going to be a lifetime project, and so it has been.

Although the house is not an architectural masterpiece, it does have character. Its brown-shingle exterior gives it an affinity with the Craftsman houses of the area, but actually, it is a little older and has details that are distinctly Victorian.

Although I liked the character of the house, I felt no preservationist obligation to return it an "original" state. My criterion for change was to be appropriate. Nothing was sacred or indispensable—except the redwood. The doors, casings, trim, and baseboards are all redwood, as are the eight columns that surround the porch and the open staircase in the front hall.

Built When the Forests Would Last Forever

When the house was built in 1896, people were still logging the redwoods in the East Bay Hills above Berkeley and Oakland. These redwoods were huge from the evidence of the stumps that have been discovered in recent years—some of the biggest anywhere. Redwood was abundant and cheap, and its widespread use established a tradition that many architects today are understandably loath to give up.

Some architects in the past, notably Bernard Maybeck, used redwood with taste, sensitivity, and an awareness of its structural qualities as well as its beauty, but much of it was treated merely as a readily available wood and even cut into 2x4s for framing. Some mills worked nothing but redwood, producing doors, sash, and water tanks.

The popular taste of the time favored dark wood, and even though redwood is naturally dark, it was often painted or, as in the case of our house, finished with a dark,

The finished paneling extends the detailing of the adjacent entryway into the dining room. Baseboards and door casings have been stripped of paint and refinished.

lusterless shellac, to look like mahogany. When tastes changed in the 1940s and '50s, it was painted over in varying shades of white. Upstairs, we stayed with paint, and downstairs, I selected my projects carefully.

I refinished the living-room trim, the staircase, and the porch columns, but the dining room presented me with a quandary. The low chair railing and the tongue-and-groove paneling below it never appealed to me. I didn't want to put a lot of effort into stripping the paint and still have something I didn't like. Various ideas were dismissed as too ambitious or inappropriate, so we continued to live with the room exactly as it was when we moved in, bland wallpaper and all. Finally, I decided to copy the coffered paneling in the front hall. Extend it, so to speak, into the dining room.

Justifying Old-Growth Materials

This plan involved using redwood. And not just any redwood would do. It had to be old growth, preferably vertical grain, and my conscience just wouldn't allow it. Even though over the years I had used a lot on my own and other people's projects, for about ten years, I've avoided using old-growth redwood and talked other people out of using it.

About the time that the dining-room project was stirring in my mind again, I got a job building a couple of mammoth doors for a ski cabin. The architect brought me a pile of 3-in.-thick redwood planks that had been the bottom of an old water tank. Here was the solution I had been looking for. Sure, I had seen ads in *Fine Homebuilding* for recycled lumber, but I had visualized industrial timbers with nail and bolt holes and other scars of their use; not what I wanted in my dining room.

I made a trip to the source of the old tank, Recycled Lumberworks in Ukiah, Cali-

fornia, to inspect some redwood wine-tank staves. These staves had come from the Taylor Winery in Hammondsport, New York, and, of course, originally came from California. Joe Garnero, the owner, let me pick out vertical-grain staves, and I ended up buying seven of them at $2.25 per bd. ft.

The staves were 13 ft. 3 in. long and made from 3x10 stock. After flattening the concave side and squaring one edge, I re-sawed them on my 14-in. Ryobi bandsaw. This process yielded three ¾-in. boards from each stave with at least one good face on each piece. Some of the interior sides had a rich cherry-red color that penetrated about 1 in. The interior wood was, for the most part, clean and even-colored, although there was some mineral staining. Unlike the water-tank bottoms, these staves had no rot (they also gave off a pleasant aroma with each cut).

A Coffered-Panel Design

The paneling that I was copying is 5 ft. high, and consists of 10-in.-sq. panels bordered by 4¾-in.-wide frames and 1¾-in.-wide moldings. The frames are in the same plane as the plaster, with the panels inset slightly. At the top of the paneling is a 1-in. by 1-in. ogee molding set atop a 2-in. by ½-in. strip of wood with a bead along the bottom. The same detail appears on top of all the doors and windows on the first floor. There is evidence that there had once been a dentil below the ogee, but none had survived. The baseboards are 1x10 rustic shiplap with a cap molding.

Until I removed the plaster and the old wainscoting and could see the back of the front-hall wall, I wasn't sure how the paneling in the hall had been done. It turned out that the panels were set between the rough framing so that their faces were flush with the edge of the studs. The ¾-in. frame members were then nailed to the studs, which made them flush with the plaster above. For this to work, the stud spacing had to correspond to the width of the panels. My panels were predetermined in width, and the dining-room studs didn't match that width. So I needed a different approach.

I decided to make the panels ⅜ in. thick and overlay them with ¾-in.-thick frames. That gave me an assembly that was 1⅛ in. thick. With this starting place, I chose to make the wall above the paneling out of ½-in. drywall over ⅝-in. plywood to match the paneling thickness.

I drew up elevations of each wall of the dining room and altered the spacing of frames and panels as needed to match the quirky spacing of the doors, built-in cabinets, windows, and fireplace. I wanted to keep the panels as close to 10-in. square as possible.

As preparation for the new wall finishes, the author stripped the lath and plaster to reveal the original framing. Here, the new layer of ½-in. drywall has been installed over a layer of ⅝-in. plywood, for a wall thickness of 1⅛ in.

An 8-in.-wide furring strip of ⅜-in. plywood abuts the underside of the drywall/plywood sandwich, where it is nailed to the faces of the studs. Individual 1x nailers are affixed to the back of the plywood with drywall screws.

Getting Down to Business

I began by marking a point 5 ft. above the floor to match the hall wainscoting and ran a level line all around the room. I dropped it 1½ in. so that my cap molding would overlap the drywall by that much. I then nailed on the plywood and the drywall (top photo).

I milled the staves into 1x frames and ⅜-in. panels at my shop. By a stroke of luck, the frame lengths were almost exactly one-fourth of my stave lengths, so there was no waste. I got one frame width and two molding widths out of each board. I ran the moldings on my shaper in two passes with two sets of knives that I ground myself to match the original.

I established the bottom and top lines of the panels and nailed ⅜-in. furring strips to the studs. After marking the horizontal spacing of the panels, I screwed nailers to the backside of the plywood (photo, right). Now I had a nail base for the redwood panels and plywood filler strips that fall between them.

As Karpilow uses his brad gun to attach thin redwood panels to their nailers, the color and pattern of the redwood start affecting the room.

Laying Out the Frame

The largest wall section was 8 ft. long. I laid out the frame on the floor using a spacer block to represent the horizontal pieces (photos, facing page). When I was satisfied with the fit, I cut all the horizontal pieces and positioned them using vertical spacers. I then marked each joint for two biscuits and marked the pieces with chalk on the back to keep them in order and slotted them for the biscuits. Working right to left with the upper horizontal in place, I put in the uprights and short sections, nailing them as I went

Plywood strips between redwood panels provide backing for the frame and trim pieces.

With the uprights in place, the horizontal members are positioned and marked for slotting with a biscuit cutter.

Spacers help to speed an accurate layout. This framework of redwood 1x4s will overlay the wall panels. Here, the author uses spacer blocks cut the same length to mark the positions of upright frame members.

With the help of a makeshift bench on a pair of sawhorses, the author plows grooves in the sides of the rails with a biscuit cutter. A clamped-on block backs up the cuts.

along. I put the bottom piece in last, wedging it tight with blocks from the floor (photo, facing page).

Incidentally, I built this long panel in place because it was too big to prefabricate. For smaller frames, I used the same layout procedure, but instead of assembling them on the wall, I clamped them with bar clamps. This way, I not only got tighter joints but also could sand them better.

Horizontal members, glued to their left upright with biscuits, are adjusted to their pencil marks prior to being joined with the top rail and the upright already in place. Chalk marks on the backs of the pieces guard against mix-ups.

A little judicious tapping with a hammer and block fine-tunes the pieces to their registration marks.

One-by blocks between the bottom rail and the floor act as wedges to drive the rails and the uprights together ensuring tight glue joints.

Moldings Cover Up the Rough Edges

After the frames were in place, I was ready for the moldings. I sanded them in lengths using a sanding block I made for that purpose (bottom photo). I cut all the horizontal pieces first because I knew these pieces were unvarying. I knew I would likely have some variation in the verticals because on-site assemblies are never as accurate as shop-built ones, so I cut a sample piece, marked where it fit, and cut the necessary parts. Going up or down $\frac{1}{16}$ in. with my adjustable miter stop gave me a tight fit all around. I pressed the moldings into place, then brad-nailed them to the panels.

Strip the Paint and Apply the Varnish

I removed most of the baseboards and casings for refinishing. One hundred years ago, they were surfaced on only one face from stock a full 1 in. thick. So even after running it through the planer and removing the paint, as well as a century's worth of nicks and bruises, I still had material thicker than what is on the market today. I cleaned up the most exposed edges by running them through the table saw. Then I dressed the bevels with a scraper and a rabbet plane.

In some parts of the room, it was impractical to remove the old moldings. So I used a heat gun and scraper to remove paint and shellac. Then I went over the moldings with scrapers made from old bandsaw blades filed to match their profiles. This tedious exercise included a lot of sanding and filling of old nail holes. When I was done, I wondered if I should just have run all new molding.

I had planned to finish the wainscoting with a waterborne clear finish. But the tests that I made first didn't make me happy. The depth and color of the wood were practically quenched by the dull water-based varnish. I ended up using Zar Satin Quick-drying Polyurethane Varnish.

After he cut them tight to the frame members, Karpilow taps the mitered moldings into place. A couple of 18-ga. brads secure each molding piece to its panel.

We chose a William Morris wallpaper for the upper part of the wall. This was how the room should have been done in the first place. In that case, however, somebody probably would have painted it white in the '50s. And instead of writing an article, I would have been stripping paint.

Miles Karpilow makes carved doors and furniture in Oakland, California.

Traditional Cabinetry from a Modern Material

■ BY JAY GOLDMAN

A year after I installed a fireplace surround that included large, solid-wood raised panels, I was called back to replace a frame that had blown apart when its panel expanded in summer humidity.

When I made the surround, I had left room for the poplar panels to expand. But there wasn't enough expansion space for the largest panel, measuring about 3 ft. by 6 ft. It was too late in the game to change the surround's design, so I made a replacement panel out of medium-density fiberboard (MDF), which is far more dimensionally stable than solid wood. Since then, the job has been problem-free.

After that callback, I decided to change my production methods to develop an efficient system for making paint-grade raised-panel cabinets from MDF that are less susceptible to changes in humidity and temperature than solid wood. I've now used the system on several cabinet jobs with fine results.

By milling one-piece face frames and frame-and-panel components out of MDF, I can save the considerable labor required for traditional stile-and-rail joinery. In some cases, I use the same piece of MDF both for the frame and for the raised panel. The cutout, or offcut, from the frame becomes the blank for the panel.

As the price of solid wood increases and high-quality stock becomes scarce, man-made alternatives such as MDF are becoming more attractive. As long as the project is to be painted, MDF has several advantages over solid wood. MDF is generally cheaper, always available, and perfectly flat.

Dimensional stability is just one reason why MDF is ideal for raised panels. Because the material is manufactured in 4-ft. by 8-ft. sheets, it can be used to make large panels that don't have to be glued up from separate boards. It also mills cleanly. With no grain, there is little or no tearout.

Strategic Uses for MDF

The author makes cabinets out of ¾-in. shop-grade plywood, but uses medium-density fiberboard (MDF) for raised panels and even for some face frames.

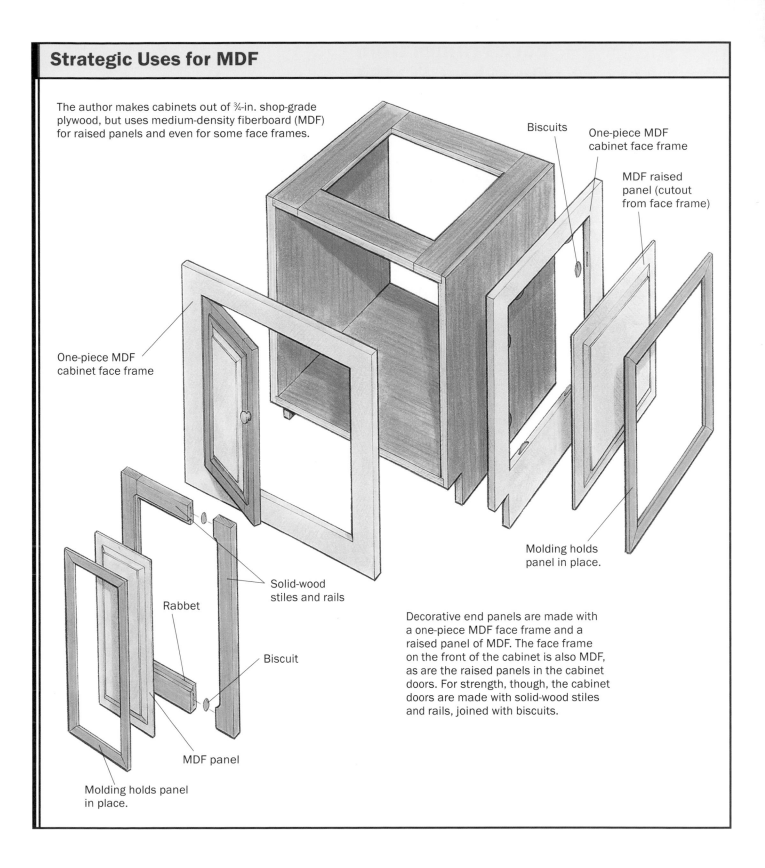

Biscuits

One-piece MDF cabinet face frame

MDF raised panel (cutout from face frame)

One-piece MDF cabinet face frame

Solid-wood stiles and rails

Rabbet

Biscuit

Molding holds panel in place.

MDF panel

Molding holds panel in place.

Decorative end panels are made with a one-piece MDF face frame and a raised panel of MDF. The face frame on the front of the cabinet is also MDF, as are the raised panels in the cabinet doors. For strength, though, the cabinet doors are made with solid-wood stiles and rails, joined with biscuits.

Face Frames from One Sheet of MDF

The first step in making the cabinet face frames is cutting the MDF to the overall outside dimension of the frame. Then I mark the outline of the stiles and rails in pencil. I cut these lines with a small circular saw, finishing up the corners with a jigsaw.

The edges of the face frames have to be clean and smooth because they will be clearly visible when the cabinet door is opened. I clean these edges up with a router and a bearing-guided, flush-cutting bit. The bearing on the bit rides against straightedges that I tack to the face of the face frame. The rounded inside corners produced by the router bit can be squared off with a file. Once the first face frame has been completed, it can be used as a template for other cabinets of the same size.

Two Ways to Make a Frame and Panel

I make frame-and-raised-panel assemblies in one of two ways, depending on how they will be used (drawing, p. 137). If the assembly is for the end panel, or return, on a run of cabinets, I make the raised panel and the surrounding stiles and rails all from MDF. If the frame-and-panel assembly will be used as a door, then I make the raised panel from MDF and the stiles and rails from solid wood.

It's possible and cost-effective to make one-piece doors by routing details into a solid sheet of MDF (sidebar, pp. 140–141). But if I'm going to the trouble of making a true frame-and-panel door, I make the stiles and rails out of solid wood, which is stronger.

When the plans call for a frame-and-panel return on the side of a cabinet, I build the cabinet box first and then overlay the sides with the panel detailing. This may sound like extra work, but I find it much easier than building a raised panel into the side of the cabinet itself.

I make a stile and rail to go around a raised panel the same way I make a cabinet face frame. I cut the stiles and rails out of a solid sheet of MDF with a circular saw and a jigsaw (top left photo, facing page).

In the case of the side panels, these saw-cuts do not have to be perfectly smooth because the edges will be covered by molding that will bridge the gap between the face frame and the panel. Also, the frame-and-panel assembly will not be visible from the back. The assembly can be nailed, screwed, or biscuited and glued to the side of the cabinet carcase.

Raised Panels Are Trapped between Biscuits and Molding

To mill raised panels for cabinet sides or for doors, I use a Williams & Hussey molder/planer, which is like a thickness planer and a shaper combined in one machine. The panels also could be milled with an ordinary shaper or router table.

Remember that it's safer to run large-diameter panel-raising bits at a lower rpm than smaller bits. On a large umbrella-shaped bit, the outside edges spin faster than the edges closer to the shaft. If your router does not have built-in speed control, you can buy a variable-speed switch from a woodworking-supply house. Also, make sure you have bulletproof dust collection because the dust produced from milling MDF is not to be believed.

On the sides of cabinets, raised panels are held proud of the face frame by biscuits glued into slots along the inside edge of the frame (photos top right and bottom, facing page). The panel is trapped in the frame by the biscuits at the back and by

To create a decorative end panel, the author makes a face frame by cutting a rectangle from a piece of medium-density fiberboard.

The raised panel will be supported by biscuits inserted in the edge of the face frame. The biscuits hold the field of the panel proud of the frame.

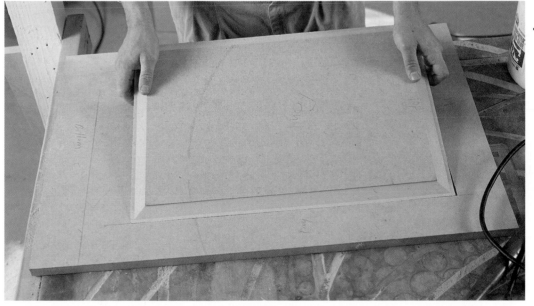

A cutout from the face frame becomes the raised panel. The author used a combination molder/planer to raise the panel. It will be held in place by moldings nailed into the face frame.

TIP

It's safer to run large-diameter panel-raising bits at a lower rpm than smaller bits. If your router does not have built-in speed control, you can buy a variable-speed switch from a woodworking-supply house.

Making raised-panel doors out of a solid piece of MDF can save labor and materials. Basically, a groove and then details such as beads, coves, and shoulders are routed into the face of the MDF sheet. For the effect to be convincing, the milling operation has to leave the simulated raised panel and frame with square or nearly square inside corners. Also, the cuts in the face of the sheet have to be made with more than one routing operation. If only one pass is made, then the details to the left and right of the groove between frame and panel will be the same.

I've made solid-MDF doors by tacking straightedges on all four sides of a blank MDF panel and then performing three different routing operations on the panel.

The first operation, done with a ¾-in.-dia. straight bit, hogs out a ⅜-in.-deep groove in the face of the panel (top photo, facing page). The next bit mills a cove on the

Most of the components that make up this cabinet, including the raised panels for the doors, drawers, countertop, and cabinet sides, were made from MDF. Once painted, the material is difficult to distinguish from solid wood.

Routers turn MDF into cabinet doors (right). Straight-edge boards hold the panel in place and serve as guides for the routers.

Sandpaper smooths the routed profile. The groove in the door has nearly square corners made with a ¼-in.-dia. flush-cutting router bit.

inside edge of the groove. Finally, a ¼-in.-dia. bit cleans up the outside edge of the main groove. This small bit nearly squares off the corners of the main groove (photo, above).

For all three operations, straightedges are left in the same place around the frame. The path of the router bits is varied with different-size bases on separate routers.

I used to have trouble with sawdust collecting inside the straightedge frame.

Part of the problem was that I was using square router bases. They are quicker to make than round ones, and also more likely to be thrown off course by accumulated sawdust. I also improved my dust-collection system. Now I use this method exclusively to make doors with a painted finish. The resulting door has a heavy feel to it, paints well, and is stable.

Pine trim covers the gap between the frame and the panel. The cabinet's panel is trapped between the trim on the front and the biscuits on the back.

Sources

Williams & Hussey Machine Co., Inc.
Riverview Mill
27 Souhegan St.
P.O. Box 1149
Wilton, NH 03086
(800) 258-1380
www.williamsnhussey.com

Three Ways to Trap the Raised Panel

Detailed below are three ways to retain a raised panel in a face frame. The top two details use biscuits to support the back of the panel and would be used only where the back of the panel doesn't show.

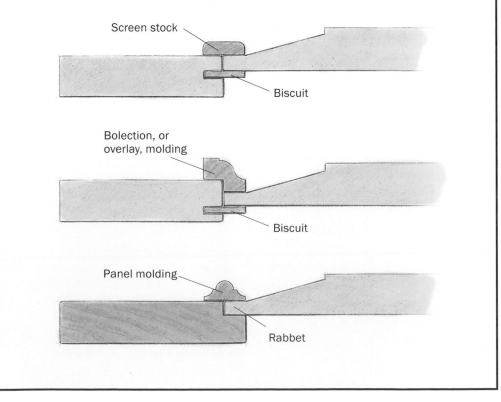

Screen stock

Biscuit

Bolection, or overlay, molding

Biscuit

Panel molding

Rabbet

bolection, or overlay, molding at the front (photo, facing page). If the edge of the panel and the face of the face frame are in different planes, then bolection molding, which is rabbeted in the back, will have to be used (drawing, facing page). Flat molding, such as screen stock, can be used if the edge of the panel and the face of the frame are in the same plane.

Because the raised side panels will be visible only from the outside, the backs of the panels do not have to be of finished quality. In fact, the panel looks quite rough from the back, with its visible biscuits and its jigsaw cuts (photo, top right).

The end result is almost impossible to tell from a traditional solid-wood raised panel and frame. A painting trick can be used to enhance this look. If the cabinets are painted with a brush, the brush strokes can be placed on the face frame to mimic the grain of solid wood.

Cabinet Doors Combine Solid Wood and MDF

For the cabinet doors, I build solid-wood stiles and rails to accept a raised MDF panel. The frame is joined with biscuits, and then rabbeted to accept the panel. The panel is held in place with bolection molding. The frame-and-panel method that I use for cabinet side panels is not suitable for doors because the doors have to be of finished quality on both inside and outside faces.

Because drawer-front frames are small, I usually make them from a solid piece of MDF. Like the door panels in the sidebar, the profiles for the drawer fronts are milled with a router.

Jay Goldman is a designer, cabinetmaker, boat builder, and teacher in Roosevelt, New Jersey.

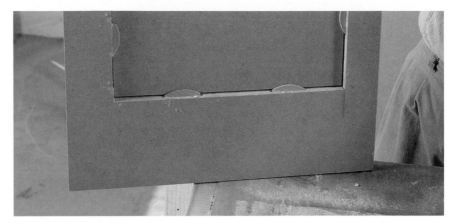

The back doesn't have to be pretty. This frame-and-panel assembly will be attached to the side of a cabinet, so its unfinished appearance on the back will be hidden.

An applied end panel is easier. Rather than integrate the MDF frame and panel into the cabinet construction, the author builds a plywood carcase and screws the MDF frame and panel to it.

Installing Elegant Wainscot Paneling

■ BY JIM CHESTNUT

This three-part system is fast to install, flexible in design, and practically unaffected by changes in humidity.

As a professional trim carpenter, I see my share of finished woodworking detail. But of all the interior trim that I do, none is more satisfying than turning out a room full of fancy wainscot paneling. Whether in entry foyers or dining rooms, painted or naturally finished, nothing adds more visual appeal for the money than nicely laid out wainscoting.

I like the look of panels made up of three basic components: a furniture-grade plywood substrate, flat stiles and rails applied over the plywood, and an overlay molding that covers the transition from plywood to stile and rail.

As shown in the drawing below, the plywood forms a continuous base in the same plane as the drywall. This method produces a large base/relief ratio. It also lends itself to tremendous variation in size and complexity without significant cost increases. For instance, you can increase the thickness of the overlay molding by ⅛ in. and its width by ¼ in. and wind up with a much bolder look

Wainscot Paneling

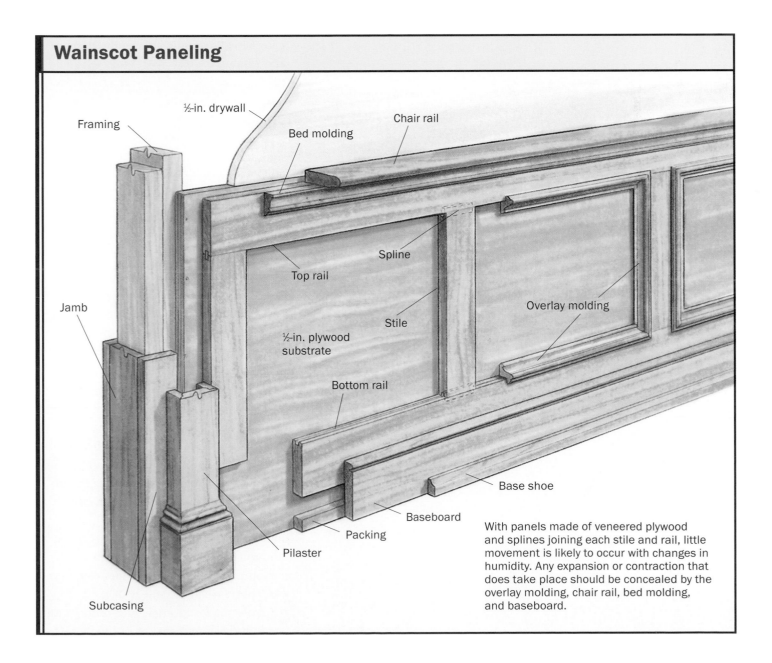

Framing

½-in. drywall

Bed molding

Chair rail

Spline

Top rail

Stile

Jamb

Overlay molding

½-in. plywood substrate

Bottom rail

Packing

Baseboard

Base shoe

Pilaster

Subcasing

With panels made of veneered plywood and splines joining each stile and rail, little movement is likely to occur with changes in humidity. Any expansion or contraction that does take place should be concealed by the overlay molding, chair rail, bed molding, and baseboard.

Replace the Drywall with Plywood

Our typical installation procedure begins with cutting out the drywall from about 31 in. down to the floor, then installing ½-in. plywood so that the plywood joints will be buried under a stile.

If the panel layout has not been predetermined, I usually sketch out what I think looks good in pencil right on the wall. This is pretty much a matter of personal preference, mine being panels that are wider than they are tall. I also am partial to tall, narrow panels in corners and adjacent to doors and windows, which sometimes enhances the layout.

Once we figure the panel layout, we install plywood, gluing and screwing scrap plywood behind joints that don't fall on studs. Our flat panel face is now in the same plane as the drywall.

The plywood goes on first. A substrate of birch veneered plywood, ½ in. thick, replaces the drywall from 31 in. down to the floor.

This style of paneling is ideally suited to job-site fabrication and is nearly immune to humidity changes.

without adding much cost. Or you can go from a simple molding to one that will add three or more shadowlines at no additional cost.

Added to its aesthetic versatility is the fact that this style of paneling is ideally suited to job-site fabrication and is nearly immune to humidity changes. Plywood is more stable than solid-wood panels. Joints between stiles and rails are glued, clamped, and reinforced with splines. And any movement that does occur is likely to be hidden by the overlay moldings.

Getting started is always the hardest part of the job (except, of course, getting a check for your start-up costs). It's a good idea to think through the ticklish details before starting, in particular the termination point of the paneling with doors and windows. More on that later.

Stiles and Rails Are Grooved for Splines

For the stiles and the rails, I use stock about ⅝ in. thick. If you take the time to flatten the stock on a jointer before finish-planing, you'll spend less time sanding the joints between stiles and rails once the material is on the wall.

After we plane, straighten, and rip our material to width, the stiles are cut to length using a stop block on the saw's fence to ensure that panels will be the same height. Then we run a groove ¼ in. wide and about ½ in. deep along one edge of the rails and along each end of the stiles. I make the grooves with a shaper, but they could be done with a router. The grooves accept ¼-in.-thick splines installed at each stile-to-rail joint. Each 1-in.-wide spline is about ¼ in. shorter than the stile width.

There are two advantages to running full-length grooves in the edges of the rails. First,

A chalkline marks the top of the paneling. The author holds a piece of chair rail in place while nailing the top rail to the plywood.

Splines reinforce the glued joint between stile and rail. Square cuts on both ends of each stile ensure that the stiles are installed plumb.

the location of a stile can be adjusted easily by sliding the stile from side to side. And second, a full-length groove leaves an escape route to blow out excess glue with the air gun to avoid glue puddling. This minimizes glue setup time and the possibility of "spline telegraphing" caused by sanding over a joint that has swollen from the absorption of water from the glue. When the wood eventually dries out and shrinks, a depression can form. I have seen glue pour out of a broken biscuit joint three days after glue-up because the cutter depth was too deep for the biscuit used.

Before doing any glue-ups, we permanently fix the top rail to the plywood (top left photo). We run a bead of yellow glue on the back of the rail close to its bottom edge to force seasonal movement away from the stile-to-rail joints. If the rail expands or contracts, the movement is then more likely to take place at the top of the top rail, which is covered by the chair rail and the bed molding.

After the glue has set up for an hour or two, we glue in the stiles and the bottom rail. This way, we can apply clamping pressure by using blocks off the floor. This is preferable to tying up valuable floor space with stile-and-rail assemblies held together with bar clamps.

Blocks wedged between the floor and the lower edge of the bottom rail force the rails and stiles together.

At each door opening, a pilaster the same height as the paneling is glued and screwed to the plywood substrate and to the flat subcasing attached at a right angle to the jamb.

At door openings, the pilaster and the flat subcasing provide a termination point for the chair rail, bed molding, and base and shoe.

Pilasters Mark the Door Openings

Our stiles and rails will be proud of the drywall by their thickness, and any subsequent layering, such as base and shoe, chair rail, and bed molding, will add to this difference. At the floor, the bottom rail and the baseboard combined are 1¼ in. thick. Even with a thick door casing, say 1⅛ in., the rail and baseboard would stand out ⅛ in. beyond the casing.

This situation could be handled with a plinth block, but the spot where the chair rail meets the door casing presents an awkward condition as well. I solve this problem by installing a hollow pilaster in place of the casing from the floor to the top of the paneling (bottom photo).

To provide a consistent surface at the door jamb for the pilaster, chair rail, and casing, we install a flat piece of stock, or subcasing, ½ in. thick by 3 in. wide, at a right angle to the face of the door jamb (drawing, p. 145). To make room for the subcasing, ½ in. is ripped off each edge of the jamb. The drywall is cut back so that the face of the subcasing will be flush with the face of the adjacent drywall. The pilaster is held back from the inside of the jamb 1½ in. to 2 in. The base, base shoe, chair rail, and bed molding now can wrap the pilaster and die into the flat subcasing.

Where paneling meets window trim, we normally notch the chair rail over the casing and the window's stool cap over the stile. This approach works for window casings because they are less noticeable than door jambs and are often hidden by curtains.

Here the author snaps a piece of coped bed molding into place. Note how the subcasing brings the jamb flush with the plane of the drywall.

The chair-rail cap hides a locking miter. The author used a locking, or mortise-and-tenon, miter to join the sides of the pilaster.

The 2½-in.-wide chair rail covers the plywood paneling, the top rail, and the bed molding.

At window openings, the stool cap is notched to fit over the stile. Note how the window casing is rabbeted to fit over the rail and stile. The chair rail will cover the exposed rabbet.

Custom Molding Makes a Tight Fit

The author uses overlay molding milled with an out-of-square rabbet. This way, the trim fits tighter against the plywood and stile or rail.

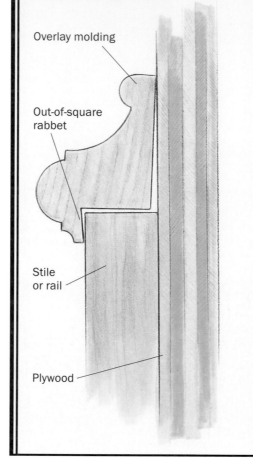

Overlay molding

Out-of-square rabbet

Stile or rail

Plywood

A Back-Cut Overlay Molding Ensures a Tight Fit

Stock overlays have a square rabbet that can leave a slight gap between the molding and the panel or the rails and stiles. I mill my own overlays so that the part of the trim in contact with the stile or rail is cut back at a slight angle (drawing, above). This ensures a tighter fit, which is especially important in stain-grade work that can't be caulked.

With this rabbet we can maintain tighter fits when the stile thickness varies due to cupping, twisting, unnoticed end snipe, or overaggressive sanding. On some jobs we've found it worthwhile to modify stock moldings by rerunning the rabbet on a router table using a dovetail bit (with featherboards) or on a table saw.

Like all moldings, overlays are subject to miss-milling and end snipe. Because of the rabbet, they are also much more delicate than most other moldings and should be handled sparingly and carefully.

After thoroughly inspecting the stock, we cut all the right-hand miters, starting with the longest pieces and working down. Then we cut the left-hand miters, using a stop affixed to the saw's fence to control the length of the cut accurately without measuring each time. Unlike the standard, square-rabbet overlay, the back-cut overlay molding is mitered while it sits on a stick that is the same thickness as the stiles and rails. This step keeps the overlay in the right position to take a miter cut.

Overlays Cut a Tad Short

The length of each piece of overlay molding can vary depending on the width of the rabbet on the back of the molding. On small moldings we usually cut the lengths so that there will be ⅛-in. minimum play both vertically and horizontally in the assembled molding within the stiles and rails (top photos, facing page). Moldings with wider rabbets can stand even more play.

I cut a few extra overlay moldings of each length in case a brad comes through the face. Extras make good mock-ups for future clients and good kindling as well.

If you're nailing the molding together by yourself, take five minutes and make a simple fixture to hold one piece rigid while you glue and shoot the next piece to it (bottom photo, facing page). If your partner is doing nothing worthwhile, make him do the holding while you do the shooting. For rapid

The overlay molding was cut to allow about ⅛ in. of play between the rabbet in the back of the molding and the stiles and rails.

Brads hold the trim in place. The molding covers the continuous groove cut in the top and bottom rails to accept the splines.

glue cleanup, we use a toothbrush and warm water. Scrub the glue with the brush; then blow the residue away with an air gun.

Before installing the overlay frames, we sometimes shoot a straightedge to the top rail with a pin tacker and run the overlay up to it, ensuring a straight installation at the top of the panel, which is the most conspicuous area. Making sure of this detail is especially important on stairway paneling, where even slight variations are easy to read.

If your budget is really limited, you can make rails and stiles out of plywood because the overlay will hide all the edges. And if your budget is really, really limited, you can substitute drywall for the plywood panels.

To keep receptacles from breaking up the look of the paneling, put them in the baseboard. It's good to have the electrician run his outlet wires long with no boxes attached and to use "old work" steel boxes mounted later.

Jim Chestnut is an interior-trim contractor and tool manufacturer who lives in Fairfield, Connecticut.

A simple fixture holds the overlay moldings square during assembly. The overlay moldings were glued and nailed together while they were clamped to fences that form a 90° angle. The fences are the same thickness as the stiles and rails.

CREDITS

p. iii: Photo by Charles Bickford, courtesy of *Fine Homebuilding,* © The Taunton Press, Inc.

Table of contents: Photos on p. iv (left) by Roe A. Osborn, courtesy of *Fine Homebuilding,* © The Taunton Press, Inc.; p. iv (right) © Marilyn Ray; p. v (left) © Jonathan F. Shafer; p. v (center) by Charles Bickford, courtesy of *Fine Homebuilding,* © The Taunton Press, Inc.; p. v (right) by Charles Miller, courtesy of *Fine Homebuilding,* © The Taunton Press, Inc.

p. 4: Basic Scribing Techniques by Jim Tolpin, issue 77. Photos on pp. 5, 6, 13, and 14 © Patrick Cudahy; p. 11 © Jim Tolpin. Illustrations © Christopher Clapp.

p. 15: Plate Joinery on the Job Site by Kevin Ireton, issue 70. Photos by Kevin Ireton, courtesy of *Fine Homebuilding,* © The Taunton Press, Inc. Illustrations by Bob Goodfellow, © The Taunton Press, Inc.

p. 22: 10 Rules for Finish Carpentry by Will Beemer, issue 113. Illustrations by Dan Thornton, © The Taunton Press, Inc.

p. 30: Pneumatic Finish Nailing by Jim Britton, issue 140. Photos on pp. 31 (left), 34 (top and bottom right), 36 (top right) and 39 (bottom) by David Ericson, courtesy of *Fine Homebuilding,* © The Taunton Press, Inc.; pp. 30 (right), 34 (lower left), 35 (top), 36 (left photos) and 39 (top) by Andy Engel, courtesy of *Fine Homebuilding,* © The Taunton Press, Inc.; pp. 35 (bottom) and 37 by Scott Phillips, courtesy of *Fine Homebuilding,* © The Taunton Press, Inc. Ilustrations by Don Mannes, © The Taunton Press, Inc.

p. 40: A Pair of Built-In Hutches by Kevin Luddy, issue 125. Photos by Roe A. Osborn, courtesy of *Fine Homebuilding,* © The Taunton Press, Inc.

p. 48: Curved Baseboard Corners by Eric Blomberg, issue 90. Illustrations by Bob La Pointe, © The Taunton Press, Inc.

p. 52: Running Baseboard Efficiently by Greg Smith, issue 76. Photos © Marilyn Ray. Illustrations by Bob Goodfellow, © The Taunton Press, Inc.

p. 57: Designing and Installing Baseboards by Joseph Beals III, issue 108. Photos by Kevin Ireton, courtesy of *Fine Homebuilding,* © The Taunton Press, Inc.

p. 68: More Than One Way to Case a Window by Joseph Beals III, issue 98. Photos by Jefferson Kolle, courtesy of *Fine Homebuilding,* © The Taunton Press, Inc. Illustrations by Dan Thornton, © The Taunton Press, Inc.

p. 78: Making Curved Casing by Jonathan Shafer, issue 67. Photos on pp. 78–81 and 83–85 by Kevin Ireton, courtesy of *Fine Homebuilding,* © The Taunton Press, Inc.; p. 82 © J. Azevedo. Illustrations © Bob La Pointe.

p. 86: Bench-Built Window Trim by Jim Britton, issue 117. Photos by Charles Miller, courtesy of *Fine Homebuilding,* © The Taunton Press, Inc. Illustration by Bob La Pointe, © The Taunton Press, Inc.

p. 94: Crown Molding Basics (originally titled Installing Crown Molding) by Tom Law, issue 51. Photos by Charles Bickford, courtesy of *Fine Homebuilding,* © The Taunton Press, Inc. Illustrations by Michael Mandarano, © The Taunton Press, Inc.

p. 102: Cutting Crown Molding by Stephen Nuding, issue 68. photos by Susan Kahn, courtesy of *Fine Homebuilding,* © The Taunton Press, Inc. Illustration © Christopher Clapp.

p. 106: Installing Two-Piece Crown by Dale F. Mosher, issue 71. Photos by Charles Miller, courtesy of *Fine Homebuilding,* © The Taunton Press, Inc.

INDEX

Index note: page references in italics indicate a photograph; references in bold indicate a drawing.